Clay Minerals: A Guide to Their X-ray Identification

Dorothy Carroll

U.S. Geological Survey, Menlo Park, California

THE GEOLOGICAL SOCIETY OF AMERICA

SPECIAL PAPER 126

Library of Congress Catalog Card Number: 71-98016

S.B.N. 8137-2126-1

Published by

THE GEOLOGICAL SOCIETY OF AMERICA, INC.

Colorado Building, P.O. Box 1719

Boulder, Colorado 80302

1970

Printed in the United States of America

The printing of this volume has been made possible
through the bequest of
Richard Alexander Fullerton Penrose, Jr.,
and is partially supported
by a grant from
The National Science Foundation.

As this volume was being completed The Society received word of Dorothy Carroll's death on January 30, 1970. The summary of much that she learned during an active and productive life will, it is hoped, stand as a fitting memorial to her contributions to the science.

Contents

ILLUSTRATIONS

FIGURE

PLATE

TABLE

Introduction

This paper is a guide to the X-ray examination of clay minerals; it incorporates background information concerning the principal crystallographic features of clay minerals, and how this is used in the X-ray identification of these minerals, together with laboratory techniques and the application of X-ray diffractometry to the diagnosis of the clay minerals in natural sedimentary materials.

Clay is a natural material of very fine texture, usually plastic when wet, and hard and compact when dry. Clay consists mainly of fine mineral particles with admixed organic matter and/or other fine mineral grains, which range in size downward from 0.002 mm (2 microns). It has a large surface area because of its fine size and the platy character of the individual minerals. The finest part of clay behaves as a colloid; fineness of grain size and other physical properties result in stickiness as well as plasticity when wet. These properties have been made use of in numerous cultural and industrial applications.

Clay is essentially a weathering product of the disintegration and chemical decomposition of igneous rocks and of some types of metamorphic rocks. In sedimentary rocks it is largely detrital, but during geologic time a certain amount of diagenesis generally takes place. In marine and nonmarine unconsolidated sediments, the clay is also at first detrital, but it may subsequently be modified by the environment. Commercial clay deposits are mostly residual, as in the Cornish kaolin deposits. Clays may be winnowed from their original environments and accumulate as pocket deposits. Those that have been derived from transportation and deposition of volcanic glass in an aqueous environment form the important bentonite beds in parts of the northern United States.

Clay, in larger or smaller quantities, is a constituent of the soils and sediments of most natural environments and of most materials that are examined by geologists and soil scientists. The easily recognizable physical properties it possesses has enabled people to use it for building (bricks) and pottery, and for the early hieroglyphic writing on tablets long before its nature was understood.

This paper describes methods for the X-ray examination and diagnosis of the clay minerals commonly found in geologic materials.

CLAY-SIZED MATERIAL

Clay-sized material is considered here to be less than 0.002 mm or $<2\ \mu$ effective spherical diameter (U.S. Department of Agriculture, 1951). There is a natural break in the continuity of weathered material at the lower limit of the silt-size, which is 2 microns. In natural material or crushed rock the larger sand-sized grains are, on microscopic examination, distinct monomineralogic grains. In gravel or coarse sand, the grains may consist of quartz by itself or of quartz and other grains. Such grains are often aggregates. To the naked eye the finest material, clay, appears to be amorphous, but although the grains are very small they are crystalline aluminosilicates.

Clay has different surface chemical reactions and different bulk physical properties from those of the larger grains of silt, sand, and gravel of weathered rocks. Whatever its ultimate composition, the fine particles are called clay.

There are many classifications of the grain sizes in natural materials, but most agree that clay is <2 microns (0.002 mm) in grain diameter. A useful chart correlating the various grain-size definitions of sedimentary materials has been compiled by Truesdall and Varnes (1950).

Soil scientists formulated an international classification of the sizes of the mineral grains in soils (U.S. Department of Agriculture, 1951, p. 207). Most examinations of the grain size distribution in soils refer to these sizes unless otherwise stated. The grain sizes in the international classification are:

Coarse sand	2–0.2 mm	Silt	0.02–0.002 mm
Fine sand	0.2–0.02 mm	Clay	<0.002 mm

CLAY MINERALS

The crystalline nature of clay minerals was long anticipated by mineralogists (Grim, 1953, p. 11-25), but it was not until 1930 that Hendricks and Fry (1930) of the U.S. Department of Agriculture were able to prove by X-ray powder photographs that clay consists of crystalline mineral particles. Even though the fact of crystallinity was known, it was not until after World War II that wide-range goniometers used on X-ray units provided the means for the rapid identification of the minerals present in clays. As a result of the pioneering efforts of many researchers, the identification of clay minerals in all kinds of sedimentary materials has become almost routine in many universities, government institutions, research institutes, and private industry.

The principal minerals in clays are layer-lattice silicates (phyllosilicates), except the Palygorskite-Sepiolite Group which are chain silicates. The layer-lattice silicates are made up of combinations of two structural units, a silicon-oxygen tetrahedron, and an aluminum-oxygen-hydroxyl octahedron. A report on the classification of the phyllosilicate clay minerals by the International Mineralogical Association was given by Brindley (1966), and the complete Report of the Nomenclature Committee (1966-67) of the Clay Minerals Society was published by Brindley and others (1968). Table 1 gives the tabulation of the Nomenclature Committee. It includes some minerals that have not been identified in clays and omits important details of others. The clay minerals commonly found in each of the major groups, kaolinite, mica, montmorillonite, chlorite, and palygorskite-sepiolite, and in clay fractions of sedimentary materials, are further classified in the description of each group. An important omission in Table 1 is the mention of the mixed-layer minerals that are so prominent in most natural situations. The position of illite has not yet been clarified by the I.M.A. Many minerals so identified may be interstratified. Mackenzie and Mitchell (1966) also published a classification of the clay minerals based on the I.M.A. classification. Their classification includes the chain silicates palygorskite and sepiolite.

OTHER MINERALS OF CLAY SIZE

Many non-clay minerals may be present within the size range of clay minerals. The presence of a mineral depends principally on its presence in the distributive province, its stability on weathering and, for some minerals, the geochemical changes that take place

TABLE 1. CLASSIFICATION OF THE CLAY MINERALS

Type	Group (x = charge per formula unit)	Subgroup	Species*
1:1	Kaolinite-serpentine $x \sim 0$	Kaolinites	Kaolinite, halloysite
		Serpentines	Chrysotile, lizardite, antigorite
2:1	Pyrophyllite-talc $x \sim 0$	Pyrophyllites	Pyrophyllite
		Talcs	Talc
	Smectite *or* Montmorillonite-saponite $x \sim 0{\cdot}25$–$0{\cdot}6$	Dioctahedral smectites *or* Montmorillonites	Montmorillonite, beidellite, nontronite
		Trioctahedral smectites *or* Saponites	Saponite, hectorite, sauconite
	Vermiculite $x \sim 0{\cdot}6$–$0{\cdot}9$	Dioctahedral vermiculites	Dioctahedral vermiculite
		Trioctahedral vermiculites	Trioctahedral vermiculite
	Mica† $x \sim 1$	Dioctahedral micas	Muscovite, paragonite
		Trioctahedral micas	Biotite, phlogopite
	Brittle mica $x \sim 2$	Dioctahedral brittle micas	Margarite
		Trioctahedral brittle micas	Clintonite
2:1:1	Chlorite x variable	Dioctahedral chlorites 4-5 oct. cations per formula unit)	
		Trioctahedral chlorites (5-6 oct. cations per formula unit)	Pennine, clinochlore, prochlorite

*Only a few examples are given

†The status of *illite* (or *hydromica*), *sericite*, etc. must be left open at present because it is not clear whether or at what level they would enter the Table; many materials so designated may be interstratified.

After Brindley and others, 1968, p. 323.

after deposition. The same factors apply to residual deposits of clay. Stability on weathering will allow for transportation and deposition of a mineral in the formation of a sediment, and for the presence of a mineral in residuum or slightly transported residual deposits. Table 2 gives the principal non-clay minerals to be expected in the clay sizes, that is, <2 microns.

Other minerals may be present under certain conditions. In addition to the minerals listed in Table 2, Brown (1961) gives titanium oxides, pyrophyllite and talc, sulfides, sulfates, phosphates, and gibbsite. Their presence can be found by X-ray diffraction only if they occur in excess of about 5 percent by volume or weight. Specific tests combined with X-ray diffraction may indicate their presence.

TABLE 2. NON-CLAY MINERALS IN THE CLAY FRACTION

Mineral	Distributive province	Residual deposit	Transported sedimentary material	Diagenetic
Quartz	+	+	+	......
Feldspar	+	?	+*	+**
Dolomite	+	......	......	+
Calcite	+	......	+***	+**
Iron oxides	+	+	+	?
Zeolites	+	......	......	+

+Present.
......Absent.
? Possibly present, depending on climate.
*Depends on type of weathering and transportation.
**Feldspar is authigenic under certain conditions; for example, in some limestones. Calcite is authigenic in some clay deposits and in soils that are not leached, such as Pedocals.
***Generally in the form of foraminiferal tests and shell fragments.

*Compiled by Dorothy Carroll

ENVIRONMENT AND OCCURRENCE OF CLAY MINERALS

There are two principal kinds of situations in which clay minerals are found:

(1) Clays are produced by alteration of minerals as rocks

weather *in situ* as a result of chemical leaching that is caused mainly by rainwater, ground water, and drainage.

(2) The altered original minerals may be removed by erosion and redeposited.

Although the end products of weathering under leaching conditions are the same in any well-drained situation, the time required for clay mineral production varies. Generally, time is measured geologically.

Clay minerals are formed as the result of chemical alteration of primary rock-forming minerals: feldspars, some ferromagnesian minerals (olivine, pyroxenes, amphiboles, mica). These minerals vary in their stability or ease of alteration; for example, the stability of the ferromagnesian minerals is as follows:

muscovite >>> amphibole (igneous) > pyroxenes >> olivine > biotite.

Hence, muscovite (and its derivative illite) is of widespread occurrence in sediments and residual deposits, whereas the other ferromagnesian minerals (with the exception of metamorphic types of amphibole) are of local derivation, and their presence denotes a short period of weathering.

The geochemical environment in which these minerals are found will modify their composition and structure in ways that are not yet completely understood. Different types of clay minerals develop from alumino-silicate parents, depending first on the weathering environment. The initial alteration is generally modified by its later geochemical environment. Thus, feldspar commonly alters first to montmorillonite because the environment of initial alteration is alkaline. In a dry climate, montmorillonite will be stable (as it is in the Wyoming bentonite beds), but with chemical leaching, an accompaniment of a moist climate and good drainage, the initial montmorillonite will change first to halloysite ($4H_2O$) and then to kaolinite. With further leaching continued for a long time, the kaolinite loses silica, and gibbsite results.

Kinds of Clay Minerals

KAOLINITE GROUP MINERALS

The minerals in this group are hydrated alumino-silicates of general chemical composition $Al_2O_3 : SiO_2 : H_2O = 1 : 2 : 2$. Structurally they are phyllosilicates of type 1 : 1 (Table 1). The individual sheets are unsymmetrical with SiO_2-tetrahedra on one side and Al_2O_3-octahedra on the other. The hydroxyl groups of the octahedral sheet are in contact with the basal oxygens of the tetrahedra of the next sheet. The kaolinite minerals thus have two surfaces, each of which is different in chemical and structural composition. There is no isomorphous replacement of cations in the structure, and there is no charge on the unit cell.

The kaolinite group minerals can be divided into two subgroups, *a*, structurally and chemically similar to kaolinite, and *b*, similar to *a* but with additional water.

a	*b*
Kaolinite	Allophane (amorphous to X-rays; contains adsorbed water)
Disordered and poorly crystalline kaolinite	
Dickite	
Nacrite	
Metahalloysite (7A)	Halloysite (10A; contains interlayer water)
Disordered halloysite	

Kaolinite

Chemical Composition. 2 $Al_2Si_2O_5(OH)_4$ or 2 $SiO_2 \cdot Al_2O_3 \cdot 2H_2O$ per unit cell. There are no isomorphous substitutions in kaolinite. The water in kaolinite exists as hydroxyl groups.

Structure. Kaolinite consists of sheets of SiO_2 tetrahedra bonded to sheets of Al_2O_3 octahedra that are continuous in the *a* and *b* directions and stacked one above the other in the *c* direction.

Unit Cell. $a = 5.14$ Å; $b = 8.93$ Å; $c = 7.37$ Å. Kaolinite is triclinic, and Z = 1, $\alpha = 91° 48'$; $\beta = 104° 30'$; $\gamma = 90°$. Each layer has a thickness of about 7.15 Å (tetrahedral sheet = 2.1 Å; octahedral sheet = 5.05 Å) which is 001 for kaolinite.

Diffraction Pattern—Oriented Mount. An integral series of *d* reflections (spacings) based on (001), that is, 7.15 Å = (001); 3.57 Å = (002); 2.35 Å = (003); 1.79 Å = (004). In most mounts of kaolinite only the (001) and (002) reflections are sufficiently intense for recognition in diffractometer charts. The intensity of the (003) and (004) reflections is insufficient for recognition in mixtures of minerals. The appearance of the (001) reflection of well-crystallized and poorly crystallized kaolinite in an oriented mount is shown in Figure 1.

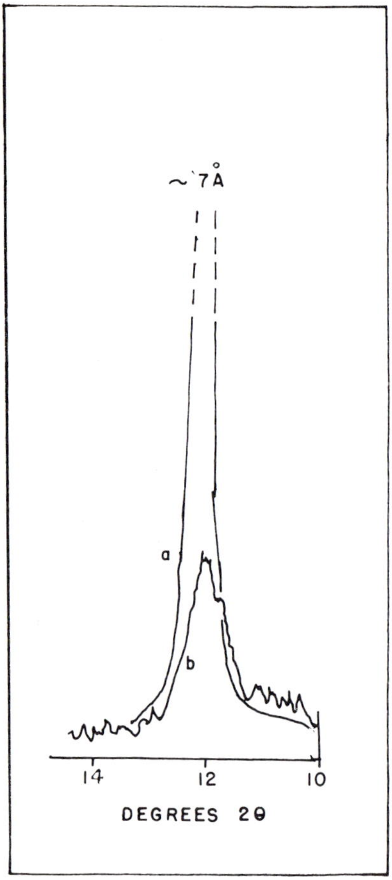

Figure 1. X-ray diffractograms of the (001) spacing of well-crystallized and poorly crystallized kaolinites (CuKα radiation, 10^3 cps., multiplier 1, ratemeter 4; chart speed 30″/hr., scanning speed 1° 2θ min., kv 50, ma 20). (a) Kaolinite, A.P.I. H-5, Bath, South Carolina. (b) Kaolinite, soil, Virginia.

Unoriented Mount. In well-crystallized kaolinite a series of reflections between 3.37 and 2.55 Å caused by *hkls* (111)-(201) and *hk* bands are characteristic of poorly crystallized kaolinite. The (060) reflection at 1.48 Å-49 Å, giving the *b* dimension of the unit cell, is usually easily obtained.

Identification. The presence of ~7 Å *d* reflection (001) and 3.57 Å-58 Å *d* reflection (002) in oriented mount; collapse of structure to an X-ray amorphous mineral, metakaolin, on heating to ~550° C; (060) *d* reflection of about 1.49 Å-1.50 Å.

Occurrence. Found in decomposed igneous rocks, in leached soils, in residual clay deposits, in hydrothermally altered rocks of suitable composition. Kaolinite is generally more abundant in regions of tropical weathering than in more temperate zones. Kaolinite occurs in detrital deposits derived from tropical weathering.

Kaolinite, Disordered

Chemical Composition. Similar to that of kaolinite. In addition to impurities in the samples there may be some isomorphous replacement in the octahedral layer.

Structure. The structure is similar to that of kaolinite but is disordered principally by random layer displacements parallel to the *b* axis. When only partial order exists, the triclinic angle $\alpha = 91.6°$ of well-crystallized kaolinite becomes 90° in disordered forms which are then pseudo-monoclinic. Brindley (1961b, p. 64-65) gives a detailed description of disorder in kaolinite. Hinckley (1963) investigated the crystallinity of the hard and soft kaolins in Georgia and South Carolina. "Hard" and "soft" describe the ease of pulverizing the two principal types of kaolin found in commercial kaolin deposits. The hard kaolin was sedimented in a marine environment, and the original detrital kaolinite was not recrystallized. The soft kaolin was sedimented in a fresh-water environment, and the detrital kaolinite was leached and recrystallized. He used the $1\bar{1}0$ and $11\bar{1}$ *d* reflections of kaolins at 20.4° and 21.3° 2θ (CuKα radiation) to measure the crystallinity. The method of measurement is shown in Figure 2, and details of the method are given by Bates and Hinckley (1959).

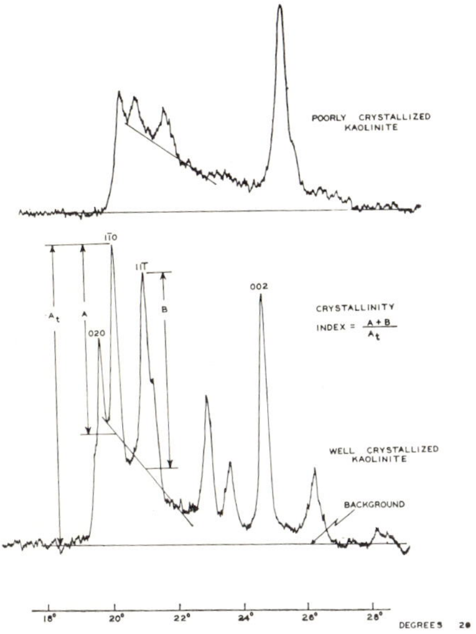

Figure 2. Crystallinity index for kaolinite (unoriented mount, *after* Hinckley, 1963). Reprinted by courtesy of Pergamon Press, Inc.

Unit Cell. $a = 5.15$ Å; $b = 8.93$ Å; $c = 7.39$ Å; $\beta = 104.5°$, and disordered kaolinite is practically monoclinic.

Diffraction Pattern. Disorder in kaolinite results in the broadening of the (001) and (002) *d* reflections, and consequently a lowering of intensity of reflections in oriented mounts. In unoriented mounts, a broad reflection in the vicinity of 4.48 Å (19.8° 2θ) occurs instead of a number of sharp reflections.

Identification. In oriented mounts the presence of disordered kaolinite is shown by broad (001) and (002) *d* reflections at approximately the same distance 2θ as those in ordered kaolinite. Figure 2 shows the appearance of ordered and disordered kaolinite in an oriented mount.

Occurrence. There is a considerable variation between well-crystallized and poorly crystallized kaolinite in the situations in which kaolinite occurs. Geologically, old kaolinitic soils may contain well-crystallized kaolinite, but the type often found is similar to that of Figure 1 (b). Clay deposits may contain both hard and soft kaolinite. The kaolin clays of the southeastern United States were investigated, among others, by Hinckley (1963), who showed that the hard kaolinites have better crystallinity than the soft kaolinites. A method for distinguishing between well-crystallized and poorly crystallized kaolinite by X-ray diffraction, known as the crystallinity index, was worked out by Hinckley (1963). The essential features of the crystallinity index are given in Figure 2.

Dickite

Chemical Composition. 2 $SiO_2 \cdot Al_2O_3 \cdot 2H_2O$, a polytype of kaolinite.

Structure. Dickite is extremely well-crystallized in individual monoclinic plates that consist of SiO_2 tetrahedra layers bonded to Al_2O_3 octahedral layers, as in kaolinite. Dickite differs from kaolinite in the stacking of the layers. Further structural details are given in Grim (1953), Brindley (1961b), and Bailey (1963).

Unit Cell. a = 5.15 Å; b = 8.95 Å; c = 14.42 Å; β = 96°48′. The unit cell is similar to that of kaolinite except that it contains two sets of SiO_2-Al_2O_3 layers.

Diffraction Pattern. Dickite gives more intense *d* reflections and a clearer diffractogram than most kaolinites in both oriented and unoriented mounts.

Identification. The diffraction pattern is similar to that of kaolinite, but the spacings give more intense diffraction than with kaolinite. The *d* reflection at about 7.16 Å is the (002) reflection; the (001) reflection at about 14 Å is not resolved. The (004) *d* reflection at 3.57 Å-58 Å is in a similar position to the (002) of kaolinite. Other reflections in the integral series based on d (002) are weak (Newnham, 1961; Newnham and Brindley, 1956).

Occurrence. Dickite was first identified as a member of the kaolinite group from material of hydrothermal origin, but later it was recognized as authigenic in many sandstones where it had apparently formed by alteration of detrital minerals and redistribution of their silica and alumina. Such material has recrystallized as dickite. Dickite is relatively common in sandstones. In the Hawkesbury Sandstone (Triassic) of the Sydney Basin, Australia, Bayliss and others (1965) have recorded the widespread presence of dickite associated with kaolinite and illite. This well-crystallized dickite developed in the sandstone authigenically, probably as a result of leaching of detrital material in a porous, although confined, situation. Dickite does not form in clayey sediments or in soils.

Nacrite

Chemical Composition. Similar to kaolinite.

Structure. Nacrite is a polytype of kaolinite. Bailey (1963, p. 1205) found that the smallest unit cell is of monoclinic shape and contains two kaolin layers, whereas Hendricks (1938) had described nacrite as a six-layer structure with β close to 90°. Two alternate two-layer unit cells exist, as in dickite, with approximately 100° and 114° (Bailey, 1963, p. 1205).

Unit Cell. a = 8.96 Å; b = 5.15 Å; c = 14.35 Å. X = b = 5.1 Å; Y = a = 8.9 Å, the reverse of the orientation in kaolinite and dickite.

Diffraction Pattern. The X-ray reflections from nacrite are weaker than those of kaolinite and dickite, and consist partly of broad bands caused by the coalescence of diffractions from several adjacent planes. The patterns are ill-defined.

Identification. The first basal spacing is similar to that of kaolinite at about 7.18 Å but is d (002); a spacing at 3.58 Å almost coincides with kaolinite (002) but is nacrite (004). An indexed powder pattern is given by Bailey (1963, p. 1206).

Occurrence. Nacrite is rare in comparison with kaolinite and dickite, but as it is difficult to identify, it may have been described as kaolinite. Nacrite has been identified as segregations of authigenic origin in altered rocks and it has also been recorded in clays associated with sedimentary rocks. Nacrite probably does not occur except in restricted situations, as does dickite (Brindley, 1961b, p. 82).

Halloysite

Chemical Composition. Similar in composition to kaolinite.

Structure. A polymorph of kaolinite (Brindley, 1961b; Bailey, 1963). The hydrated form, halloysite, consists of a regular sequence of kaolinite layers separated by interlayer water. The dehydrated form, metahalloysite, contains hydroxyl water. Halloysite dehydrates irreversibly to metahalloysite.

Unit Cell. $a = 5.15$ Å; $b = 8.93$ Å; $c = 7.37$ Å or 10.27 Å, the difference in the c parameter being due to a single molecular sheet of water.

Diffraction Pattern. Halloysite dried at 110°C has basal spacings (001) and (002) similar to kaolinite. Prior to dehydration the mineral is either in the $4H_2O$ form or near to it, and the 7 Å d spacing is replaced by one at about 10.1 Å. The basal spacings are broad, implying an irregular stacking sequence. In unoriented mounts, broad reflections occur at the position of (*hko*) reflections.

Identification. By less sharp d basal reflections than kaolinite, and by the collapse of the 10.1 Å spacing to about 7.2 Å on dehydration. If halloysite is not completely dehydrated, it can be rehydrated to a 10.1 Å spacing. Halloysite may dry out, so that the humidity of the atmosphere should be controlled to prevent its reversion to metahalloysite. The morphology of most halloysite is tubular as seen in electron micrographs (Bates and others, 1950).

Occurrence. Halloysite is formed in the weathering of igneous rocks in the sequence montmorillonite → halloysite → metahalloysite → kaolinite (Carroll and Hathaway, 1963). Moist conditions are necessary for the stability of halloysite. Halloysite has been recorded from Pacific Ocean sediments by Griffin and Goldberg (1963, p. 734), and it is probably more abundant in marine sediments than has so far been recorded.

Halloysite, Disordered

There appears to be a gradation between kaolinite and halloysite. Halloysite, and hence its dehydrated form, metahalloysite, has a less ordered structure than kaolinite. Halloysite consists of a highly random stacking of kaolinite layers (further details of the relation of the water layers and layer stacking are given in Brindley, 1961b). Detailed structure analyses cannot be made of halloysites because of their structural disorder.

Anauxite

It has been shown recently that anauxite is a form of kaolinite. The excess silica reported in early analyses is probably amorphous silica that may have been deposited between kaolinite platelets. The Nomenclature Committee of the Clay Minerals Society has ruled that anauxite is not a distant species and that the name should be discontinued (Brindley and others, 1968, p. 324).

Allophane

Chemical Composition. Hydrous alumino-silicate; a mixture of SiO_2 and Al_2O_3 of variable composition ranging from 1 : 1 to 1 : 2, with n H_2O. The alumina may be either in 6-coordination or in 4-coordination. The water is adsorbed.

Structure. It was thought that allophane had no regular arrangement of atoms in its structure, but soil allophane is now known to have different degrees of order (Wada, 1967, p. 690).

Unit Cell. No definite lattice parameters.

Diffraction Pattern. Although largely amorphous to X-rays, Wada (1967, p. 691) has shown that several broad X-ray reflections may occur. The strongest reflections are in the vicinity of 13.0-18 Å, 7.6-7.8 Å, 3.26 Å.

Identification. Allophane is not identifiable by X-ray diffraction alone and is therefore similar to amorphous clay, much of which may actually be allophane. Infrared absorptions give absorption in the region $800\text{-}1300^{cm^{-1}}$, but varies with the development of order in the structure produced by aging. D.T.A. curves show a gradual loss of water up to about 300°C; some apparently more ordered structures show the endothermic reaction between 400° C and 525°C suggestive of loss of hydroxyl water. Fieldes and others (1966, p. 623) suggest testing for allophane by reaction with NaF to release hydroxyl groups.

Occurrence. Allophane is produced in the first stages of alteration of ferromagnesian minerals in basic and ultrabasic rocks by incongruent solution of the primary minerals, from weathering of volcanic glass, by weathering of feldspar under conditions where sericite does not form, by precipitation of hydrous alumino-silicates under low temperature and pressure in fissures in ore veins, and on continued grinding of muscovite and micaceous minerals. Allophane is probably present in many situations where it has not been recorded because of difficulty in identification. Soil allophane has been described in detail by Fieldes and others (1966).

Amorphous Clay

Amorphous clay should be identified as allophane, because a sample containing amorphous clay gives the same type of X-ray diffraction pattern as allophane. The X-ray diffractogram of a sample containing amorphous clay (allophane) can be assessed by the fact that the same volume of such clay as that containing well-crystallized clay minerals gives a background radiation without, or with only very weak spacings from which individual minerals cannot be identified with certainty. The greater the dilution of crystalline clay minerals with amorphous clay, the weaker the spacings of minerals and the greater the background radiation.

MICA GROUP MINERALS

Minerals with a 2 : 1 layer structure can be arranged in a logical sequence as follows:

Mica

- Muscovite (dioctahedral)
  - Partial replacement of K^+ ions with H_2O and metallic cations.
    - Illite
    - Glauconite
    - Celadonite
    - Biotite (trioctahedral)
    - Vermiculite
- No interlayer K^+ ions, but H_2O and metallic cations are present.
  - Montmorillonite Group

The chemical relationships among members of the dioctahedral micas are shown in Figure 3.

The characteristics of the micaceous clay minerals have been summarized by MacEwan (1961, p. 145), who tabulated the main features as given in Table 3.

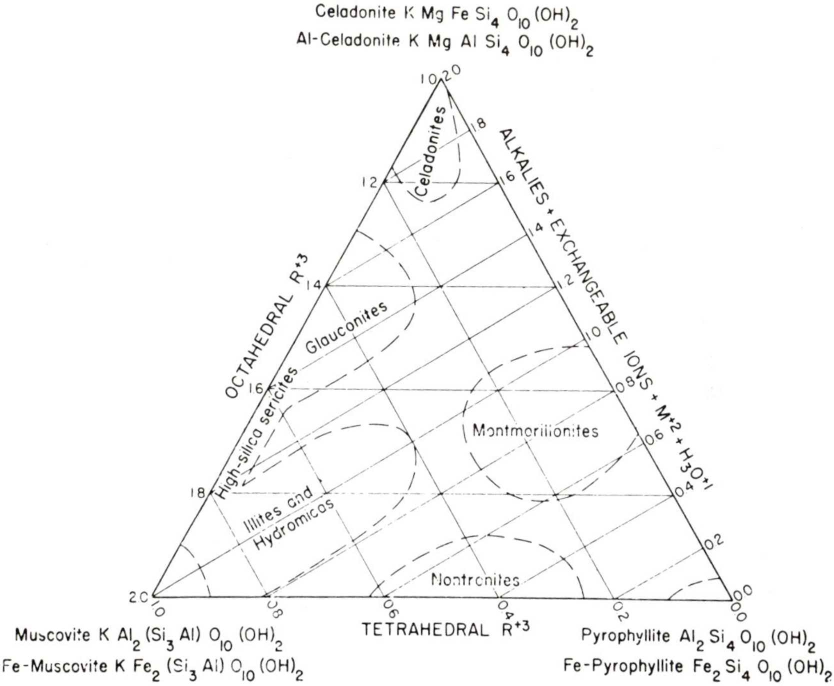

Figure 3. Mineralogical relations of members of the dioctahedral mica group (*after* Yoder and Eugster, 1955). Reprinted by courtesy of Pergamon Press, Inc.

Muscovite

The most ubiquitous micaceous minerals in soils, sedimentary rocks, and unconsolidated deposits of mineral matter are related to muscovite. Mineralogically, micas are classified as muscovite, a dioctahedral mineral of composition $KAl_2(AlSi_3O_{10})$, or as biotite, a trioctahedral mineral of composition $K(Mg,Fe)_3O_{10}(OH)_2$. Muscovite is much less altered by weathering than biotite. Within the muscovite group of minerals there is much isomorphous replacement.

Chemical Composition. The members of the dioctahedral potassium mica group vary between $KAl\ Al(Si_3Al)O_{10}(OH)_2$ and $KAl\ Mg(Si_4)O_{10}(OH)_2$.

Structure. Muscovite and the micaceous clay minerals consist structurally of one layer of alumina octahedra sandwiched between two layers of silica tetrahedra. This structure is generally referred

TABLE 3. CHARACTERISTICS OF MICACEOUS CLAY MINERALS*

Mineral type	Charge per unit-cell-layer	Main interlamellar cations	Swelling behavior
Brittle mica	4	Ca	No swelling reported
Mica	2	K, Na	May swell slightly with special treatment
Vermiculite	1·3	Mg	Many organic complexes
			May swell indefinitely with special treatment
Montmorillonite	0·7	Na, Ca, etc.	Wide range of complexes
			Swells indefinitely with some monovalent cations
Pyrophyllite-talc	0		No swelling reported

**After* MacEwan, 1961. (Reprinted by courtesy of The Mineralogical Society.)

to as a 2 : 1 layer lattice (Table 1). The tripartite layers are bonded together by K^+ ions in 12 coordination with oxygens of the silica tetrahedra. The packing of the oxygen ions in the ideal muscovite structure leads to offsets of upper and lower hexagonal nets (at the base of the silica tetrahedra) in any of the three 120° directions. Consequently, polymorphs or polytypes of muscovite are formed. These have been designated as 1 M, 2 M_1, 2 M_2, and 3 T, for one-layer monoclinic muscovite, two kinds of two-layer monoclinic muscovite, and three-layer trigonal muscovite by Smith and Yoder (1956). Details of the structural arrangements of the common polytypes are given by Yoder and Eugster (1955). Table 4 gives the principal *d* reflections of the polytypes of muscovite as shown by X-ray powder photographs using CuK_α radiation.

The muscovite polytypes produced by different structural arrangements can be recognized by the (*hkl*) reflections shown in diffractograms of unoriented mounts of clays. In particular, the (112) reflection (22.375° 2 θ) and the 112 (24.315° 2 θ) are useful for diagnosis, using CuK_α. In metamorphism, the disordered structure (1 Md) is gradually replaced by a more ordered structure and *hkl's* begin to appear; the intensity of the *hkl* reflections may be used as a measure of the amount of change from disordered mica to the 1 M polytype. With increasing temperature and pressure, the 1 M structure changes to that of the 2 M structure. The transformations are slow. Both the 1 Md and the 1 M muscovite polytypes are metastable (Velde, 1965, p. 448).

The muscovite polytypes commonly found in the clay-sized fraction of sedimentary materials are given in Table 5.

Muscovite grains in natural situations are almost exclusively the 2 M polytype because mica has a very slow rate of inversion; that is, mica adjusts very slowly to the equilibrium conditions of the chemical environment. Because of this, 2 M muscovite is detrital in sediments. As weathering proceeds, the 2 M polytype will slowly change its composition by the replacement of some K^+ ions by H_2O. Apparently, this replacement is limited because even experimentally the spacing of (002) cannot be increased much beyond 12 Å.

In addition to the mica polytypes described above, detailed single crystal determinations have shown that many others can be recognized (Radslovich and Norrish, 1962; Ross and others, 1966). Such polytypes cannot be identified except by single crystal X-ray studies.

Dioctahedral micas which have either 1 Md or 1 M polytypes are either metastable muscovite forms or are micas with a compo-

TABLE 4. X-RAY *d* REFLECTIONS FROM 8° TO 35.2° 2θ FOR 2M, 1M, AND 1 Md SYNTHETIC MUSCOVITE POLYTYPES WITH CuKα RADIATION

(Compiled from data of Yoder and Eugster, 1955)

2M			1M			1Md		
hkl	Å	I	*hkl*	Å	I	*hkl*	Å	I*
002	10.014	<100	001	10.077	<100	001	10.077	
004	5.021	55	002	5.036	37	002	5.036	
110	4.478	55						
111	4.458	65	020	4.488	90	020	4.488	
021	4.391	14	$11\bar{1}$	4.349	27	003	3.356	
111	4.296	21	021	4.115	16	130	2.582	
022	4.109	14	$1\overline{12}$	3.660	60	200	2.550	
112	3.973	12	003 022	3.35	<100			
$11\bar{3}$	3.899	14	112	3.073	50			
023	3.735	32	113	2.929	6			
$11\bar{4}$	3.500	44	023	2.269	16			
006 024	3.351	<100	130	2.582	50			
114	3.208	47						
025	2.999	47						
115	2.871	35						
116	2.803	22						

*Intensities not given by Yoder and Eugster (1955).
(*See* relation of illite and muscovite in Figure 4, p. 20.)

TABLE 5. MUSCOVITE POLYTYPES PRESENT IN THE CLAY-SIZED FRACTIONS OF SEDIMENTARY MATERIALS

(Compiled from data of Yoder and Eugster, 1955)

Polytype	Structure	Stability	X-ray identification	Natural environment
1 Md	Disordered 1 M	Low temperatures, <200°C; <15,000 psi.	Basal *d* spacings only; (060) at about 1.50Å; (02) band at about 4.1Å.	Common in sediments, soils; probably diagenetic in marine sediments; forms mixed-layer minerals. Illites
1 M	0° stacking angle of layers	Low temperatures, 200-250°C; <15,000 psi.	Basal *d* spacings and *hkl's* for 11$\bar{1}$, 021, 11$\bar{2}$, 022 and 024, 112, 11$\bar{3}$, 130 1$\bar{3}$1, 200, etc., (Table 4)	Sediments; low-grade metamorphic rocks to chlorite grade; secondary alteration zones; low-temperature pegmatites; in mixed-layer structures with montmorillonites.
2 M	Continuous alternation of 120° and 240°, or of 60° and 300° stacking angles	High temperatures, >200°-350° at 15,000 psi (upper limit of stability is 625°C at this pressure).	Basal *d* spacings and numerous *hkl* reflections (Table 4.)	Igneous and high-grade metamorphic rocks. Detrital in sediments and soils.

sition differing from that of muscovite; for example, glauconite, celadonite, and illite. Polytypes 1 Md and 1 M are common in nature and form a large part of the minerals in sedimentary and low-grade metamorphic rocks. The very abundant 1 Md and 1 M dioctahedral muscovites in sediments are probably authigenic, whereas the 2 M muscovites are detrital.

Unit Cell. $a = 5.3$ Å; $b = 9.2$ Å or 9.3 Å (muscovite and most lithium-bearing micas have b near 9.0 Å; the Fe- and Mg-muscovites have b 9.2 Å to 9.3 Å); $c =$ about 2θ Å; $\beta = 95°$-$100°$.

Diffraction Pattern. An oriented mount of muscovite gives an integral series $00l$ spacings of which the (001) for 2 M structures (the 002) is prominent at about 10 Å. A reflection at about 5 Å is also readily observed, but a reflection at 3.3 Å merges with that of quartz. In an unoriented mount, *hk* reflections are numerous and sharp in well-crystallized muscovites.

Identification. The mica structure is easily recognized by an intense reflection at about 10 Å and a small reflection at about 5 Å. Polytypes can be recognized in unoriented mounts from the information in Tables 4 and 5.

Occurrence. The common micaceous mineral of soils and sedimentary materials.

Illite

Illite is a field term that is used for the mica in soils, unconsolidated sediments, and sedimentary rocks. Hower and Mowatt (1966) investigated the chemical composition and structure together with a number of other characteristics of type samples of illites, K-bentonites, and mixed-layer mica-montmorillonites and found that all the illites they examined were of the 1 Md polytype.

Chemical Composition. Similar to that of muscovite but contains less K^+ ion and more SiO_2 and H_2O than muscovite. The charge due to the depletion of the K^+ ion is partially replaced by metallic cations that may be exchangeable in part.

Structure. 1 Md muscovite; but as illites represent detrital material as well as diagenetic, 2 M muscovite is found in some illites. The described illites contain any or all muscovite polytypes, and most illites are mixed-layer structures. Some samples of Fithian illite consist of mixed-layer 1 Md + montmorillonite + 2 M muscovite (Yoder and Eugster, 1955, p. 249).

Diffraction Pattern. As illite generally is a 1 Md muscovite, only the (001) *d* reflections are present (Table 3). The appearance of the principal (001) reflections of muscovite and illite are shown in Figure 4.

Identification. The diffraction pattern is similar to that of a 1 Md muscovite. Some expansion with glycol is to be expected if the mineral is interlayered with montmorillonite.

Occurrence. The mica commonly found in soils and unconsolidated sediments is mostly illite, except near a source of 2 M muscovite.

Glauconite

Chemical Composition. Similar to that of illite, but with greater iron content. The Fe^{+3} to Fe^{+2} ratio increases with oxidation.

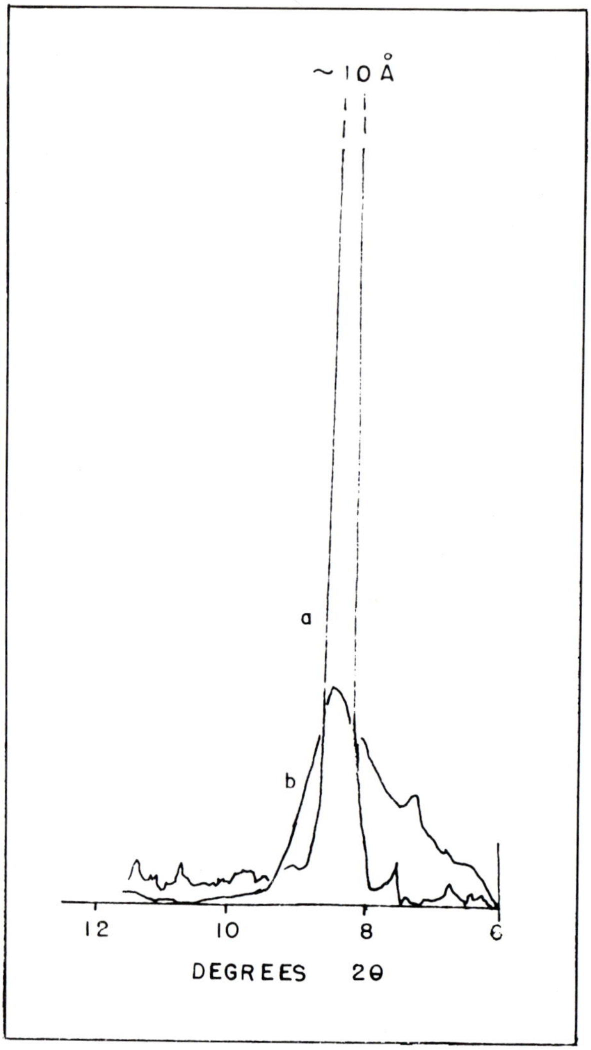

Figure 4. X-ray diffractograms of the (001) spacing of muscovite and illite (Cu$K\alpha$ radiation, 10^3 cps., multiplier 1, ratemeter 4; chart speed 30″/hr.; scanning speed 1° 2θ min., kv 50, Ma 20). (a) Muscovite, Upton County, Georgia (U.S. National Museum no. 105013). (b) Illite, A.P.I. H-35, Fithian, Illinois (calcite and quartz removed).

Structure. 1 Md and 1 M muscovite.

Unit Cell. Similar to that of 1 M muscovite.

Diffraction Pattern. Similar to that of illite. Newly formed glauconite has 1 Md reflections; older glauconite in consolidated sediments that have been subjected to pressure have 1 M reflections, and the amount of change from 1 Md to 1 M is shown by the development of *hk* reflections. The type glauconite was described from sedimentary beds in the Paris Basin, France; it has a 1 Md diffraction pattern. A green color, indicating ferrous iron, suggests reducing conditions under which glauconite has formed. Experimentally, illite will not develop the green color of Fe^{+2} in a reducing environment, but nontronite does so (Carroll, 1958, p. 10). Further information concerning glauconite is obtainable from Warshaw (1957), Burst (1958), Hower (1961), and Hendricks and Ross (1941).

Occurrence. Glauconite forms authigenically in a marine environment with Eh 0 to −150 mv. and pH 7-8. The method of authigenesis probably is: Montmorillonite (nontronite) in reducing conditions → nontronite + K (Ca, Mg) → glauconite (Md) + pressure and aging → glauconite (1 M) (Carroll, unpub. data).

Celadonite

Chemical Composition. $KAlMg(Si_4)\ O_{10}(OH)_2$. Isomorphous replacement in the octahedral layer causes a charge deficiency.

Structure. 1 M dioctahedral muscovite.

Unit Cell. Similar to 1 M muscovite.

Diffraction Pattern. As for 1 M muscovite.

Identification. By chemical analysis. Celadonite requires special conditions for formation and is not detrital.

Occurrence. An alteration product of olivine in vesicular basalt. Its magnesium content is derived from olivine, and aluminum is not an essential constituent.

Biotite

Chemical Composition. Biotite is a trioctahedral mica whose composition varies between Mg-rich (phlogopite) and Fe-rich (siderophyllite and lepidomelane) end members. A general formula for biotite is:

$$K(Mg,Fe)_3\ (AlSi_3O_{10})\ (OH)_2.$$

Structure. Lamellar monoclinic plates with irregular outlines; formed of one layer of alumina octahedra sandwiched between two layers of silica tetrahedra as in muscovite. The K^+ ion bonds the layers together.

Unit Cell. $a = 5.31$ Å; $b = 9.23$ Å; $c = 20.36$ Å.

Diffraction Pattern. Very similar to that of muscovite, but *d* spacings differ from those of muscovite because of unit cell dimensions. *d* (060) is 1.52 Å-53 Å. (Brown, 1961, p. 239)

Identification. As biotite is trioctahedral *d* (060) is 1.52 Å-53 Å compared with muscovite (dioctahedral, *d* (060) 1.50). The intensity of the reflections of the basal spacings decreases with the Fe-content, and are always less than those of muscovite because of absorption by iron when CUKα radiation is used.

Occurrence. Biotite is unstable in weathering and only occurs near its source in igneous or metamorphic rocks. It may be common in glacial sediments derived from biotitic rocks. Biotite is detrital in volcanic ash.

Vermiculite

Chemical Composition. The general composition is $Mg_3Si_4O_{10}(OH)_2 \times H_2O$, but isomorphous replacement is common in both the octahedral and tetrahedral layers. Vermiculite is commonly interlayered with biotite or with chlorite (Hendricks and Jefferson, 1938). Commercial vermiculite is interlayered unaltered biotite with vermiculite; the structural formulas obtainable from the chemical analyses of such mixtures do not indicate the composition of vermiculite alone. One structural formula calculated for vermiculite from Libby, Montana, is:

$$(Al_{.62}Fe^{+3}_{.04}Fe^{+2}_{.03}Mg_{2.70}Mn_{.01})\ (Si_{3.32}Al_{.67})\ O_{10}\ (OH)_2\ (K_{.20}Na_{.22}Ca_{.20})$$

Gruner's (1934, p. 559) chemical analysis of vermiculite (U.S. Natl. Museum, No.95647) from Bare Hills, Maryland., gives the following structural formula:

$$(Fe^{+3}_{.24}\ Fe^{+2}_{.04}\ Ni_{.01}\ Mg_{2.70})\ (Si_{2.73}Al_{1.26})O_{10}(OH)_2\ (Mg_{.55}Ca_{.01})$$

Structure. Å 2: 1-layer structure similar to mica but having, in addition, water molecules occupying about 5 Å as an interlayer.

Unit Cell. $a = 5.3$ Å; $b = 9.20$ Å; $c = 28.91$ Å. The thickness of the water layer varies slightly with the exchangeable cation present (Barshad, 1948).

Diffraction Pattern. In an oriented mount the lowest angle reflection is 14 + Å, which is taken as (001), and is followed by an integral series of basal spacings. Most vermiculite is trioctahedral with (060) $d = 1.50\text{-}53$, but dioctahedral vermiculite with d (060) of 1.48-49 has been recorded in extremely fine fractions of weathered muscovite (Rich and Obenshain, 1955).

Identification. Vermiculite will not expand much beyond 14 Å on glycolation. By heating to 500°C the interlayer water is expelled, but vermiculite quickly rehydrates; on heating to 700°C, vermiculite will collapse to 9.3 Å for (001) and will not rehydrate. In KCl solutions vermiculite reverts to a mica structure, has a low cation exchange capacity, and will not expand.

Figure 5 shows the effect of dehydration of vermiculite and uptake of the K^+ ion by vermiculite.

Occurrence. Vermiculite occurs principally as a hydrothermal product of biotite or phlogopite alteration. It is generally interlayered with biotite. Vermiculite occurs in soils where it is common as a mixed-layer mineral with chlorite and illite-type mica.

MONTMORILLONITE GROUP MINERALS

Clay minerals in which the component layers are not tightly bonded by K^+ ions (mica) or Mg^{+2} ions (vermiculite), but contain water molecules in a similar situation, have been grouped together as montmorillonites. These minerals, although having the same structural features, differ somewhat in chemical composition. They are all extremely fine-grained, do not form macroscopic crystals,

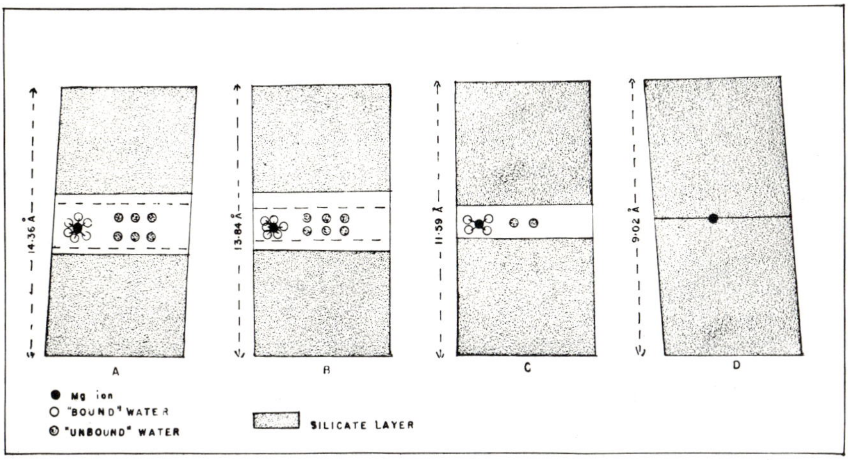

Figure 5. Dehydration of vermiculite (schematic diagram modified *after* Walker, 1961, p. 305). (A) fully hydrated; (B, C) stages in dehydration; (D) fully dehydrated.

and swell on the addition of water or organic liquids. Such minerals have variously been called smectites (British usage), montmorillonoids, or montmorins. Isomorphous substitution results in charge deficiencies which may occur either in the octahedral or tetrahedral layer. The charge deficiency is neutralized by exchangeable cations

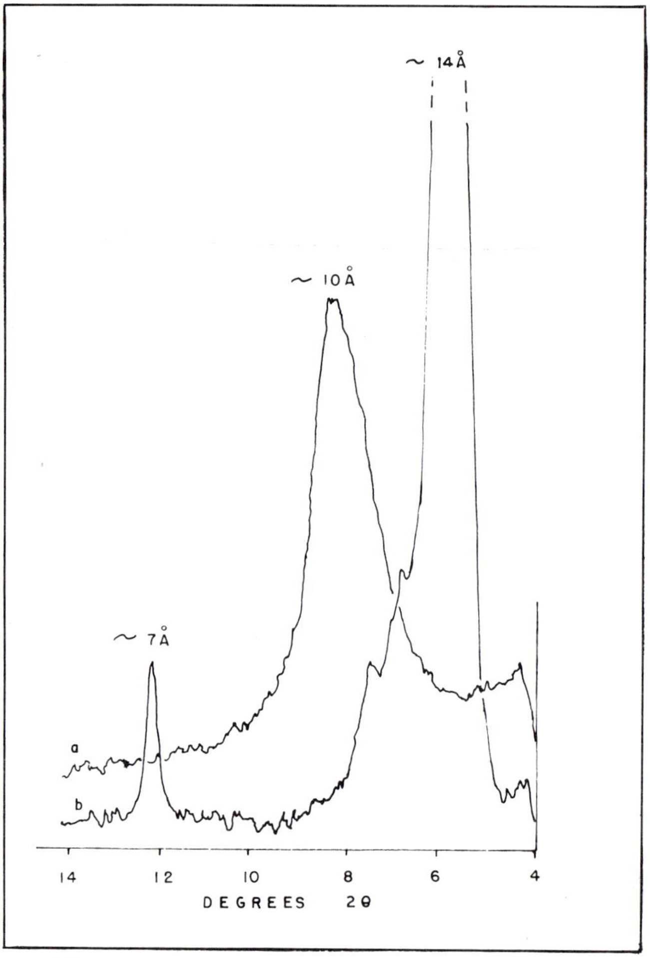

Figure 6. Potassium uptake in vermiculite from Libby, Montana. The commercial vermiculite was completely converted to the Mg-form by warming for about 48 hours in *N* $MgCl_2$ solution. (a) Mg-vermiculite treated with 1*N* KCl solution; (b) Mg-vermiculite. X-ray diffractograms of oriented mounts ($CuK\alpha$ radiation, 10^3 cps., multiplier 1, ratemeter 4; scanning speed 1° 2θ min., chart speed 30″/hr.; kv 50, ma 20 (Dorothy Carroll, unpub.).

so that the mineral becomes electrically neutral. A general scheme for the chemical composition was given by Ross and Hendricks (1945) as follows:

exchangeable ions
↑
[Ions with octahedral coordination] [Ions with tetrahedral coordination] $O_{10}[OH]_2$

A = alumina octahedron   B = silica tetrahedron

The crystal structure is BAB, with charge deficiency generally in A, but in some minerals (beidellite, nontronite, and saponite) the charge deficiency is in B.

Because of the infinite variation of possible isomorphous replacements, and consequently in the amount of charge deficiency, it is impracticable to define definite species, although a gradation between charge deficiency on A and on B occurs. Only sauconite, saponite, stevensite, hectorite, and nontronite are well-defined species. Their structural chemical formulas, calculated from chemical analyses, follow.

Sauconite	$(Zn_{2.40}Mg_{.18}Al_{.22}Fe^{+3}_{.17})(Si_{3.47}Al_{.53})O_{10}(OH)_2$
Saponite	$(Mg_{2.92}Fe^{+3}_{.08}Fe^{+2}_{.03})(Si_{3.40}Al_{.60})O_{10}(OH)_2$
Stevensite	$(Mg_{2.88}Fe^{+3}_{.02}Mn_{.02})(Si_4)O_{10}(OH)_2$
Hectorite	$(Al_{.01}Li_{.30}Mg_{2.67})(Si_4)O_{10}(OH)_2$
Nontronite	$(Fe^{+3}_{1.92}Fe^{+2}_{.05}Mg_{.10})(Si_{3.51}Al_{.49})O_{10}(OH)_2$

Most montmorillonites are dioctahedral, but hectorite, stevensite, sauconite, and saponite are trioctahedral. The commonly occurring montmorillonites are dioctahedral and belong to the montmorillonite-beidellite group, as defined by Ross and Hendricks (1945). Such montmorillonites are the principal, and generally the only, clay minerals in bentonites.

Montmorillonite

Chemical Composition. Hydrous aluminosilicate (*see* introduction to this section).

Structure. Similar to mica, but has interlayer water instead of K^+ ions between the layers; there is little bonding (Table 1) and macroscopic crystals are not formed.

Unit Cell. *a, b,* and *c* are variable within narrow limits; *a* and *b* vary with the composition of the octahedral layer, *c* with humidity and exchangeable cations. The method of describing micaceous minerals according to the occupancy of cations is due to Stevens (1946). In the montmorillonite group, both dioctahedral and trioctahedral minerals occur as listed above (p. 25). The following semiempirical formula relating the b axis in dioctahedral and trioctahedral layer silicates to the chemical composition has been suggested (Brindley and MacEwan, 1953, p. 15-59):

Dioctahedral montmorillonite, $b(\text{Å}) = 8.92 + 0.06x + 0.09y + 0.18r + 0.27s$

Trioctahedral montmorillonite, $b(\text{Å}) = 9.19 + 0.06x + 0.12p - 0.06q + 0.06s$

where p, q, r, s are the numbers of Al^{+3}, Fe^{+3}, Mg^{+2} and Fe^{+2} ions per half-unit-cell layer in octahedral coordination, and x the number of Al^{+3} ions in tetrahedral coordination. The relation of a to b is: $a : b = 1 : \sqrt{3} = 0.577$ (MacEwan, 1961, p. 168). The c spacing for most montmorillonites is in the vicinity of 15 Å.

Diffraction Pattern. Characterized by a series of reflections from the (001) plane in oriented mounts. In untreated montmorillonites the sequence of reflections is irrational, but in glycol- or glycerol-treated montmorillonites that give a uniform (001) spacing, the reflections are rational. Uniform conditions of interlayer cation (Ca is preferred), humidity, and glycolation are required to produce an integral series of spacings from (001) that can be interpreted with confidence in clays containing montmorillonite. Most mounts of montmorillonites, if not definitely unoriented, will give only the basal spacings, as shown in Figure 7.

An unoriented (random) mount may show weak *hk* reflections as given by MacEwan (1961, p. 192) and reproduced in Table 6.

The *hk* reflections are most satisfactorily recorded in Guinier powder photographs, but they can also be obtained by packing a dry-powder sample into the sample holder shown in Plate 1 and making a diffractogram. As the *hk* reflections are weak, the (02) reflection at about 4.5 Å is generally the only one found in diffractograms of unoriented mounts.

Identification. By characteristic swelling on glycolation from a *d* spacing of about 15 Å to one of about 17-18 Å. The interlayer water is removed by heating at 300°C for at least an hour, and the original 15 Å reflection is replaced by one at 9-10 Å. Characteristic diffraction patterns of the (001) spacing are given in Figure 7. It is usual to glycolate samples of clay minerals for the identification of montmorillonite because of the expansion of the (001) reflection and the sharpness obtainable from the integral series of reflections from the basal spacing. In an X-ray powder photograph of montmorillonite, both basal and general reflections are seen. The general reflections, *hk*, are characteristic of the mineral and the $00l$ basal reflections are characteristic of the condition, that is,

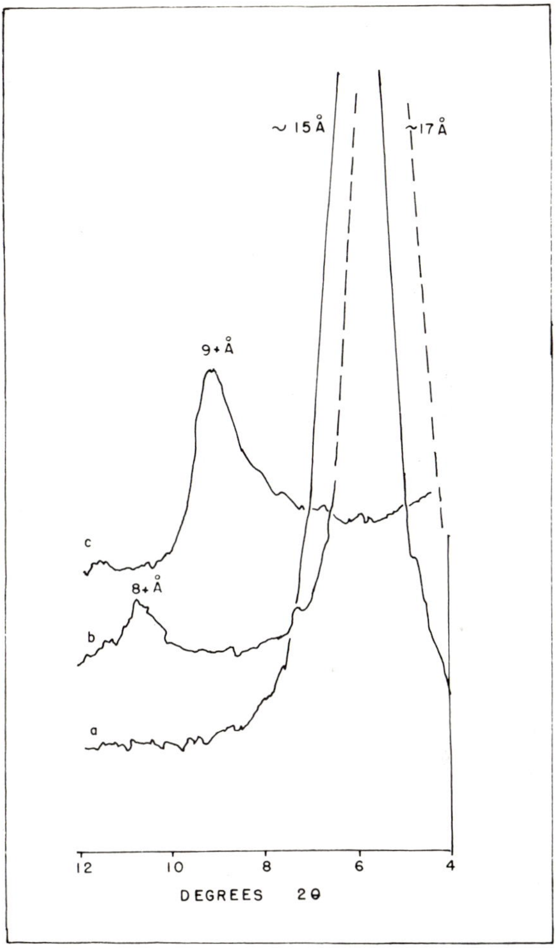

Figure 7. X-ray diffractograms of the (001) spacing of montmorillonite (oriented mount) from Chambers, Arizona (A.P.I. H-23). (a) room temperature and humidity; (b) glycolated; (c) heated 1 hr. at 300°C. CuKα radiation, 10^3 cps., multiplier 1, ratemeter 4; scanning speed 1° 2θ min., chart speed 30″/hr.; kv 50, ma 20 (Dorothy Carroll, unpub.).

interlayer water, cations in exchange positions, complexes with organic liquids (glycol or glycerol). For further details, MacEwan (1961) should be consulted.

Occurrence. Montmorillonite group minerals occur in altered volcanic ash as in bentonite beds; they are common as an alteration

Table 6. Spacings of *hk* Reflections of Montmorillonites

Indices	d (Å) calc.	1 Montmorillonite d (Å)	I	2 Montmorillonite d (Å)	I S	3 Beidellite d (Å)	I	4 Nontronite d (Å)	I	5 Volkhonskoite d (Å)	I S	6 Saponite d (Å)	I	7 Hectorite d (Å)	I	8 Sauconite d (Å)	I S	9 Pimelite d (Å)	I S	10 Cardenite d (Å)	I	11 Lembergite d (Å)	I
11; 02	4·50	4·51	10	4·61	10 vbr	4·46	10	4·56	10	4·461	3	4·52	8	4·58	10	4·62	6	4·21		4·50	7	4·56	10
												(4·07)	6						8 br	4·43	5		
												(3·02)	6					4·13		(3·04)	1		
13; 20	2·60	2·62		2·56	8 vbr	2·61		2·63		2·575	5 br	2·55	8	2·66	8	2·67		2·425	10 vbr	2·65	5	2·62	
		2·56	0				10		8			2·46	4	2·49	6		6						8
		2·42				2·50		2·57								2·584						2·47	
						2·39	4																
22; 04	2·25	2·24		2·221	3 br	2·244				2·24	½			2·289	3					2·298	1		
			3			2·168	3																
		2·159				2·094																	
		(1·909)				(1·894)	3																
31; 15; 24	1·706	1·711		1·692	6 br	1·697		1·719		1·706	½	1·74	6	1·721		1·757	3	1·691	3 br	1·739		1·70	5
			6				6		3						6						1		
		1·685				1·657		1·671						1·692						1·682			
33; 06	1·503	1·500	10	1·492	10 vsh	1·491	10	1·522	10	1·506	6	1·52	8	1·530	10	1·547	8	1·513	8 sh	1·536	6	1·536	10
										(1·483)	1									(1·483)	1		
26; 40	1·301	1·295	6	1·289	6 br	1·287	6	1·319		1·302	2 br	1·32	6	1·323		1·330	5	1·307	5 br	1·326	1	1·326	5
									3						8								
								1·303						1·304									
35; 17; 42	1·252	1·250	6	1·244	6 sh	1·244	6	1·269		1·267	½	1·29	6	1·270	2	1·288	2 br						
									3	1·254	1 br												
								1·249															
08; 44	1·127			1·120	3 sh							1·157	4			1·136	2 br						
37; 28; 51	1·036			1·029	3 br							1·052	4			1·069	2 br						
19; 53; 46	0·984	0·972	3	0·9765	4 br	0·972								0·992	4	1·008	2 br						
0,10; 55	0·902																						
39; 60	0·8683	0·865	3	0·8642	4 sh	0·864	3							0·880	4	0·893	2 br						
2,10; 48; 62																0·846	1 br						
1,11; 57; 64																0·823	2 br						
a	5·21 Å	5·18 Å		5·17 Å		5·15 Å		5·24 Å		5·18 Å		5·31 Å		5·25 Å		5·35 Å		5·24 Å		5·32 Å		5·32 Å	
b	9·02 Å	9·00 Å		8·95 Å		8·95 Å		9·13 Å		9·02 Å		9·20 Å		9·18 Å		9·28 Å		9·08 Å		9·22 Å		9·22 Å	

(1) Unter-Rupsroth, Rhön; $[Si_{7\cdot48}Al_{0\cdot52}]^{IV}[Al_{3\cdot54}Fe^{3+}_{0\cdot06}Mg_{0\cdot40}]^{VI}O_{20}(OH)_4Ca_{0\cdot32}Na_{0\cdot14}Mg_{0\cdot08}$, spacings and formula from Nagelschmidt (1938).

(2) From Wyoming bentonite; typical formula of such material, $[Si_{7\cdot76}Al_{0\cdot24}]^{IV}[Al_{3\cdot10}Fe^{3+}_{0\cdot38}Fe^{2+}_{0\cdot04}Mg_{0\cdot52}]^{VI}O_{20}(OH)_4Na_{0\cdot68}$. X-ray data by MacEwan (unpublished elsewhere).

(3) Black Jack Mine, Carson Co., Idaho; $[Si_{6\cdot92}Al_{1\cdot08}]^{IV}[Al_{3\cdot92}Fe^{3+}_{0\cdot08}]^{VI}O_{20}(OH)_4Ca_{0\cdot46}Na_{0\cdot04}Mg_{0\cdot04}K_{0\cdot02}$, spacings and formula from Nagelschmidt (1938).

(4) Behenjy, Madagascar; $[Si_{7\cdot14}Al_{0\cdot86}]^{IV}[Al_{0\cdot16}Fe^{3+}_{3\cdot68}Fe^{2+}_{0\cdot08}Mg_{0\cdot16}]^{VI}O_{20}(OH)_4Ca_{0\cdot30}Mg_{0\cdot26}$, spacings and formula from Nagelschmidt (1938).

(5) Thompsons, Utah; analysis Al_2O_3 19·70, Fe_2O_3 (incl. FeO) 3·10, Cr_2O_3 1·67, MnO 0·21, NiO 0·02, MgO 3·72, CaO 0·23, TiO_2 0·45, SiO_2 and $H_2O\pm$ not determined, spacings and analysis from McConnell (1954).

(6) Spacings from Kerr (1937), *d* values appear to be too low especially near top of table.

(7) Hector, San Bernardino Co., California; $[Si_{7\cdot78}Al_{0\cdot10}Mg_{0\cdot12}]^{IV}[Mg_{5\cdot34}Li_{0\cdot66}]^{VI}O_{20}(OH)_4Ca_{0\cdot08}Na_{0\cdot84}K_{0\cdot02}$, spacings and formula from Nagelschmidt (1938).

(8) Yankee Doodle Mine, Leadville, Colorado; $[Si_{6\cdot60}Al_{1\cdot40}]^{IV}[Zn_{3\cdot70}Mg_{0\cdot28}Al_{1\cdot58}Fe^{3+}_{0\cdot04}]^{VI}O_{20}(OH)_4Ca_{0\cdot23}Na_{0\cdot10}$; X-ray data after Faust (1951), formula after Ross (1946).

(9) Synthetic material of small particle size; formula given as $Si_8Ni_6O_{20}(OH)_4$ by van Voorthuijsen and Franzen (1951); X-ray spacings deduced from their microphotometer trace by MacEwan.

(10) Carden Wood, Aberdeenshire; $[Si_{6\cdot12}Al_{1\cdot81}]^{IV}[Al_{0\cdot88}Fe^{3+}_{1\cdot36}Fe^{2+}_{0\cdot30}Mg_{2\cdot96}Ca_{0\cdot24}]^{VI}O_{20}(OH)_4M^{+}_{0\cdot40}$, spacings and formula from MacEwan (1954).

(11) Moniwa, Japan; an iron-rich saponite $[Si_{7\cdot24}Al_{0\cdot76}]^{IV}[Al_{0\cdot08}Fe^{2+}_{2\cdot90}Mg_{3\cdot04}]^{VI}O_{20}(OH)_4Ca_{0\cdot46}$, spacings and formula from Sudo (1954). (Formula shows imperfect change balance.)

Notes on Table

(i) Indices are those for a two-dimensional ortho-hexagonal lattice $a = 5\cdot21$ Å, $b = 9\cdot02$ Å; *d* calc. gives *hk*0 spacings which are approximately coincident with band heads.

(ii) Lattice parameters of the various minerals given in last two lines are calculated from the observed spacings; for specimens 1, 3, 4, and 7 from 26,40 for *a* and 33,06 for *b*, for specimens 2, 5, 6 and 10 from the higher order reflections and adjusted to make $b = \sqrt{3}a$, for specimens 8, 9, and 11 they are obtained from 33,06 and the relation $b = \sqrt{3}a$.

(iii) When two or more spacings are given against a group of indices they probably represent subsidiary maxima or apparent edges of bands.

(iv) Reflections in brackets may be due to impurities.

(v) I = relative intensity; S = shape of band, v = very, sh = sharp, br = broad.

After MacEwen, 1961, p. 192-193. (Reprinted by courtesy of The Mineralogical Society.)

Sample holder for X-ray diffraction of unoriented clay samples (patterned after Byström-Asklund, 1966). Clay is packed normal to *b* direction of flakes. *Left,* in position for packing sample; holder is held vertical in a styrofoam base. *Right,* in position for inserting in goniometer of a Norelco diffraction unit (left side of holder is set flush with shaft of goniometer). The aluminum sample holder illustrated was made in the workshops of the U.S. Geological Survey in Washington, D.C.

SAMPLE HOLDER FOR X-RAY DIFFRACTION OF CLAY SAMPLE

product of weathering igneous and metamorphic rock minerals of suitable composition in an alkaline environment with little leaching, and are present in soils developed from suitable rocks in unleached environments (for example, in arid climates). Montmorillonite has formed authigenically from volcanic ash in the South Pacific Ocean clays (Griffin and Goldberg, 1963, p. 735), but in other oceans montmorillonite appears to have been distributed from weathered land surfaces. Detrital montmorillonite changes diagenetically to mixed-layer chlorite-vermiculite in the marine environment (Berry and Johns, 1966, p. 192).

CHLORITE GROUP MINERALS

The minerals classified as chlorites are trioctahedral phyllosilicates with a 2 : 1 layer structure similar to that of mica. The general composition of the group is $(Mg,Fe,Al)_6 (Al,Si)_4O_{10} (OH)_8$, but isomorphous replacement is usual. They consist of a 2 : 1 : 1 layer structure in which mica and brucitelike layers alternate. The mica layer is negatively charged, but is neutralized by a positive charge in the brucitelike layer which thus acts in a similar manner to the K^+ ion in muscovite. The isomorphous replacements found in analyzed specimens of chlorite have been described by Foster (1962), who classified chlorites with respect to replacement of Si by Al tetrahedral coordination, and on the Fe^{+2} : R^{+2} ratios. Chlorites fall naturally into Fe-chlorites, intermediate, and Mg-chlorites. It should be noted that Fe^{+3} is a necessary ion in many chlorites and that it is not generally due to oxidation. Figure 8 gives Foster's (1962) grouping of the chlorites.

Another very useful classification of the chlorites is that given by Hey (1954, p. 277-292). He uses the ratio $R = Fe/(Fe + Mg)$ for measuring the proportion of Fe^{+2} ions in the structure.

In addition to variation in chemical composition, Brown and Bailey (1962) have shown that structural chlorite polytypes occur. Details of the possible structural arrangements of the brucite layer, resulting in chlorite polytypes, are given by Brown and Bailey (1962), and they have recognized that 80 percent of the chlorites are the 11b polytype in which the structure is a monoclinic unit cell. Most of the remaining chlorites are the 1b polytype based on an orthohexagonal (or orthorhombic) unit cell. Two other polytypes, 1a and 11a, based on a monoclinic and an orthohexagonal unit cell, respectively, are of a limited occurrence in nature. The polytypes are due to the composition of the layers rather than to the stacking sequence of the layers. The formation of polytypes is due to equilibrium conditions as related to the energy available in the environment. The stable chlorite in normal-grade metamorphism and in medium- and high-temperature conditions is nearly always the 11b polytype. Diagenetic chlorites are the orthohexagonal 1b polytype (like chamosite) or monoclinic 1b polytype. With time and energy (for example, pressure), the 1b polytypes will recrystallize into 11b polytypes. The conversion of the 1b polytype to the 11b polytype is important in the low-grade metamorphism of iron ores as, for example, the Biwabic iron formation of Minnesota and others (Brown and Bailey, 1962, p. 845-848). The diagenetic chlorite found in marine sediments is the 1b polytype; it is poorly crystalline and probably is the variety of chlorite known as chamosite (Carroll, in prep.; *see* Fig. 8).

The 1b chlorite polytype, apart from its poor crystallinity, is difficult to distinguish from the 11b polytype, except in X-ray photographs of single crystals and in unoriented mounts of chlorite-bearing clays. Brown and Bailey (1962, p. 839) give the following *d* spacings for $20\bar{2}$, 202, $20\bar{1}$, and $20\bar{3}$ of the common polytypes of chlorite.

Polytype	Cell	*hkl*	*d*, Å	I*
11b	Monoclinic	$20\bar{2}$	2.59	5
11b	Do.	201	2.54	8
1b	Do.	$20\bar{2}$	2.61	1.5
1b	Do.	201	2.55	0.5
1b	Orthohexagonal	202	2.50	10.0
1b	Do.	203	2.33	1.0

*Measured in scale of 0-10.

Chlorite

Chemical Composition. $(Mg,Fe,Al)_6(Si,Al)_4O_{10}$ $(OH)_8$, with isomorphous replacements.

Structure. Chlorite consists of "talc" and "brucite" layers that are stacked in various ways to form polytypes as described above (Brown and Bailey, 1962, p. 819).

Unit Cell. The unit cell is slightly variable according to the chlorite polytype. The common 11b polytype has $a = 5.33$ Å; $b = 9.23$ Å; $c \sin \beta = 14.2$ Å, with $\beta = 97°$. b varies with the number of Fe^{+2} ions in the octahedral layer, and c with the number of Al^{+3} ions replacing Si^{+4} in the tetrahedral layer (Brindley, 1961c, p. 268-270).

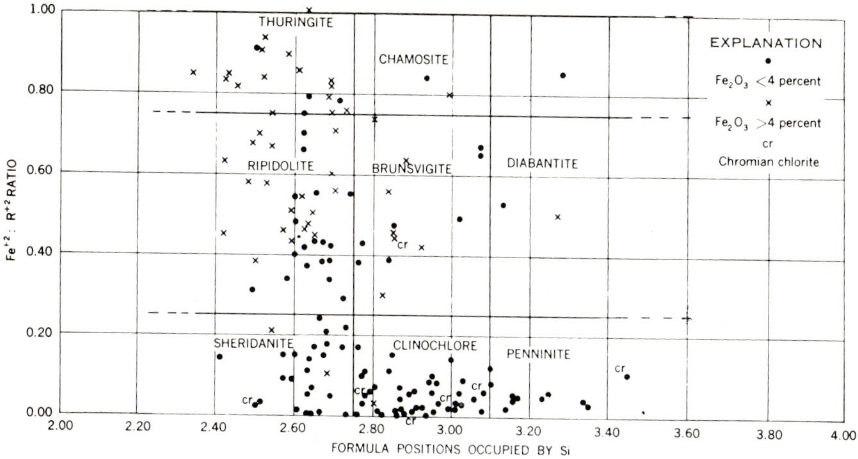

Figure 8. Classification of chlorites based on the two principal types of ionic replacement (Foster, 1962, p. A19).

Diffraction Pattern. An oriented mount of chlorite shows an integral series of (00l) spacings from 14 Å (Brindley, 1961c, p. 260-261). In an unoriented mount, many *hkl* spacings are observable, but a photographic method may be necessary to identify the polytypes.

Identification. After a chlorite diffractogram has been recognized, the following features should be investigated:

(1) Most Mg- and low Fe-chlorites give a clear sequence of the first four or five basal reflections measured from *d* (001) = 14 + Å.

(2) Fe-rich chlorites give weak 1st and 3rd basal reflections but strong 2nd and 4th order basal reflections.

(3) The *d* (001) at 14 + Å will not expand on glycolation.

(4) With warm 1 N HCl treatment, well-crystallized chlorites tend to dissolve slightly, the less ordered chlorites dissolve. Kaolinite does not dissolve with this treatment.

(5) If chlorite is heated for 1 hour at 600°C, the kaolinite structure is destroyed, and the mineral becomes amorphous to X-rays. The chlorite structure is not destroyed, and the 14 Å *d* reflection is intensified by heating up to 600°C, and the higher orders disappear.

Occurrence. The 11b polytype occurs in low-grade metamorphic schists in pre-Tertiary consolidated sedimentary rocks and metamorphosed iron-bearing sediments. It is detrital in sedimentary environments, and especially in sedimentary material derived from the glaciation of metamorphosed rocks. The 1b polytype occurs diagenetically in marine sediments and in unmetamorphosed iron ores where it is authigenic. It is also authigenic in altered ferromagnesian minerals (Brown and Bailey, 1962, p. 845).

Chlorite Varieties

Swelling Chlorite. It has been observed that some chlorite-type minerals exhibit properties intermediate between normal chlorite and montmorillonite- and vermiculite-type minerals. Some such minerals are interlayered mixtures. Long-spacing swelling minerals of chlorite-type of various structural (and probably chemical) compositions have been termed swelling chlorite. In general, minerals called swelling chlorite have a *d* spacing in the vicinity of 28 Å which increases to 32 Å on absorption of glycerol or glycol, and collapses to about 13.8 Å after heating at 450°C-500°C. However, these minerals are structurally complex and do not all behave similarly with glycerol and heat treatment. The complexities of the structure and nomenclature of swelling chlorites are described by Brindley (1961c, p. 280). Swelling chlorites are, in some respects, similar to mixed-layer minerals composed of chlorite, montmorillonite, and vermiculite.

Corrensite (named for Professor C. W. Correns of Göttingen) is such a mineral from the Keuper Clay in Germany (Lippman, 1954). It has been noted under *Vermiculite,* p. 23, that chlorite is generally associated with the alteration of biotite, and it is to be expected in hydrothermally altered biotite.

Dioctahedral Chlorite. Although most chlorite is trioctahedral, Rich and Obenshain (1955) identified dioctahedral chlorite in a soil derived from muscovite schist in Virginia. Later, Brydon and others (1961) observed a dioctahedral chlorite in a soil from British Columbia and were able to obtain a sufficiently pure sample for chemical analysis. The X-ray diffraction of this chlorite is similar to that of cookeite, a high Al-chlorite (Brown and Bailey, 1962, p. 839). The British Columbia dioctahedral chlorite has *d* (060) 1.496 Å, thus differing from trioctahedral chlorites which have *d* (060) 1.53 Å-154 Å. Brydon and others (1961), in discussing the origin of dioctahedral chlorite, point out that "chloritic" minerals may form in the A horizon of acid soils of pH 4.5-5.5 or less. Sedimentary dioctahedral chlorites (Eggleston and Bailey, 1967, p. 687) are the 1a layer type (cookeite) which are high energy structural forms that may crystallize metastably in a sedimentary environment. Dioctahedral chlorite could be authigenic.

Chamosite

Chemical Composition. Iron silicate of somewhat variable composition containing nearly 40 percent FeO (James, 1966, p. W5). Chamosite is common in many sedimentary iron ores, and its relation to chlorite is shown in Figure 8.

Unit Cell. Two types of chamosite have been recognized; one is a 2 : 1 : 1 chlorite that has long been called "chamosite." The other is a 1 : 1 type mineral which should be called "berthierine" (Brindley and others, 1968, p. 324). Two types of kaolin-like chamosite have been described by Brindley (1961a, p. 105). One of these is orthorhombic and the other is monoclinic. The cell dimensions are: *a*, 5.39 Å; *b*, 9.33 Å; $c = 7.04$ Å, with the monoclinic form having $\beta = 104.5°$. The iron in fresh chamosite is Fe^{2+}, but on oxidation the iron changes to the Fe^{3+} which causes a slight contraction of the unit cell. Brindley (1961b, p. 106) gives these figures:

	Ferrous form	Ferric form
a	5.41 Å	5.25 Å
b	9.38 Å	9.10 Å
d (001)	7.11 Å	7.06 Å

Bailey (1963, p. 1198) considers that 1 M chamosite is a trioctahedral analogue of kaolinite.

The chlorite-type chamosite has $a = 5.3$ Å; $b = 9.3$ Å; $c = 14.3$ Å, with $\beta = 97°$.

Diffraction Pattern. Brindley (1961b, p. 104) discussed the structure of chamosites, and Brindley and Youell (1953) described ferrous and ferric chamosites. In nature there may be mixtures of orthorhombic and monoclinic kaolin types and of ferrous and ferric chamosite as well as chlorite-type chamosite. In a diffractogram only the basal *d* reflections are usually shown, and the peaks may be diffuse. In a powder photograph, a trio of medium intensity lines at 2.40 Å, 2.15 Å, and 2.67 Å are diagnostic, and a band at 4.5 Å, corresponding to the (02) of disordered kaolinite, is common. Chamosite is often associated with siderite, some of the *d* spacings of which coincide with those of chamosite.

Identification. Chamosite is an authigenic mineral, generally a 1b chlorite polytype in which Fe^{+2} gives a green color which is maintained in reducing conditions and the presence of decomposing organic matter. On oxidation, the ferrous iron changes to the ferric ion, and the diffraction pattern (and structure) resembles that of a disordered kaolinite. Even when first formed, chamosite has a rather poor crystalline structure and diffractograms give only a 7 Å (001) *d* reflection. With pressure, as in the consolidation of a sediment, and low-grade metamorphism, chamosite recrystallizes to a thuringite or bavalite, in which state a 14 Å *d* reflection appears. Chamosite is then an Fe-chlorite. In powder photographs with FeKα radiation, authigenic chamosite from the Niger Delta described by Porrenga (1965) and X-rayed by Schoen and Carroll (unpub. data), has many *hk* lines that are not visible on a diffractogram. As chamosite has a 1b chlorite polytype structure, unless ferrous iron is found in appreciable quantities by chemical analysis, it would probably be identified as either disordered kaolinite or as a 1b chlorite. The (060) spacing is 1.56 Å for ferrous chamosite, and 1.514 Å for ferric chamosite (Brindley and Youell, 1953, p. 58). The kaolin-type chamosite gives X-ray data and a heating pattern like kaolinite; the chlorite-type chamosite gives X-ray data and a heating pattern like chlorite. That is, the kaolin-type structure collapses at about 550°C and appears amorphous, and the chlorite-type structure has an intensified 14 Å *d* reflection and does not collapse. However, Fe-chlorites in general are not as heat resistant as the Mg-chlorites. FeKα radiation should be used in chamosite identification (for further details on structure and identification, *see* Brindley, 1961b, p. 104).

Occurrence. Chamosite occurs in many geologically young iron ores such as the Northampton Iron Sand (Taylor, 1949) and the minette ores of Lorraine (Bubenicek, 1960). A summary of the occurrence of chamosite is given by James (1966). Chamosite occurs in marine sediments in which reducing conditions prevail, and it is probably formed diagenetically in deep-sea clays where, being unstable, it oxidizes and produces the well-known Red Muds. The X-ray diffraction pattern of such material indicates a 1b chlorite polytype rather than a kaolinite (Carroll, in prep.).

MIXED-LAYER CLAY MINERALS

Clay minerals that consist of interstratified layers of different minerals are of very common occurrence in soils and sediments.

Weaver (1956) reported that 70 percent of some 6000 sedimentary rocks examined by him contained some variety of mixed-layer clay. The fact that all clays except the Palygorskite-Sepiolite Group are similar in structure, accounts for ease of aggregation of layers of similar structure but slightly different chemical composition. Mixed layering in clays is an adjustment to environment in two ways: (1) degradation as in weathering rocks and soils; and (2) diagenesis as in deposits of detrital materials. A further step in diagenesis is in consolidation which leads to the production of more stable crystals of minerals from original detrital clay minerals. This type of very weak metamorphism leads, for example, to the formation of 2 M polytypes of mica from 1 Md illite, and to the recrystallization of glauconite into chloritelike minerals or into well-crystallized micas. MacEwan (1949) has pointed out that interstratification may be (1) regular; (2) random; or (3) a segregation within one crystallite into zones of types (1) and (2).

Because of original weathering of igneous rocks in various climates, the principal types of mixed-layer clay to be expected in the different climatic zones are:

(1) in moist tropical regions: interstratification in the sequence montmorillonite-halloysite-kaolinite (Carroll and Hathaway, 1963, p. F32);

(2) in humid temperate zones: interstratification in the sequence montmorillonite-chlorite-mica or mica-mica-intermediates-illite (Jackson and others, 1952).

The weathering of sedimentary rocks containing mixed-layer minerals will result in detrital mixed-layer clays, which may further change in the environment into which they are released by weathering.

There is no simple process of change. Changes in the chemical composition of the phyllosilicates result from the removal or addition of cations, either through the removal of K^+ ions bonding the mica sheets together, or in the stabilization of the chemical composition with cations in exchange positions that are present in the environment. The effects of anion exchange and adsorption have not been fully assessed (Carroll, 1959, p. 765-767). The final result is toward simplification of clay minerals in a leaching environment, and to preservation of the clay minerals in a neutral or nonleaching environment. In a leaching environment, simplification toward a kaolinitic type is caused by stripping of first, interlayer cations, and second, by removal of divalent cations from the octahedral layers of clay minerals. In other words, an original 2 : 1 layer structure is gradually replaced by a 1 : 1 layer silicate. The influence of natural environments was explained by Carroll (1962a, Pl. 1).

MacEwan and others (1961), explain mixed layering as a special form of intergrowths. These are classified as:

(1) Hydrated layers in clay minerals—montmorillonites and vermiculites—are hydrated varieties of mica. Mica-vermiculite and mica-montmorillonite interstratifications are common.

(2) Gibbsitic (Al_2OH_6) and brucitic $(Mg,Al)_6$ $(OH)_2$ sheets may occur between layers, giving mica-chlorite interstratifications, and with water-layers, vermiculite-chlorite, or montmorillonite-chlorite as well as interstratifications of mica-montmorillonite-chlorite. In addition, interstratification, not to be confused with mixed layering, occurs artificially as in organic complexes with montmorillonite (glycolation, for example), or by replacement of exchangeable cations (as in the replacement of Mg^{+2} in vermiculite with K^{+}). Interstratification may result from the replacement of existing exchangeable cations of clay minerals by change of chemical environment as, for example, the deposition of clay minerals from a fresh-water environment into a marine environment. Sea water has a calculable quantity of cations and anions whose activity enables them to react with clay minerals.

The possible types of mixed layering to be found in clay minerals has been calculated by Cole and Hosking (1957) as shown in Figure 9.

MacEwan and others (1961), have classified mixed layering in clay minerals statistically as regular, random, and segregation of one mineral in zones within another mineral. In regular mixed layering the diffraction pattern follows a completely regular sequence of X-ray reflections which can be interpreted in a similar way to that of the diffraction of a single mineral. Both Zen (1967) and Reynolds (1967) have described the theoretical background for the diffraction of a regular mixed-layer mineral as similar to that of a one-dimensional crystal. An example of regular mixed layering is *rectorite* (= allevardite, Brown and Weir, 1963), described by Bradley (1950), in which an alternation of pyrophyllitelike and montmorillonitelike layers give a basal spacing, *d* (001) of 25 Å. The basal spacings form an integral series from (001). Kodama (1958) described a regularly interstratified pyrophyllite-montmorillonite with a basal spacing of 25 Å which is similar to rectorite.

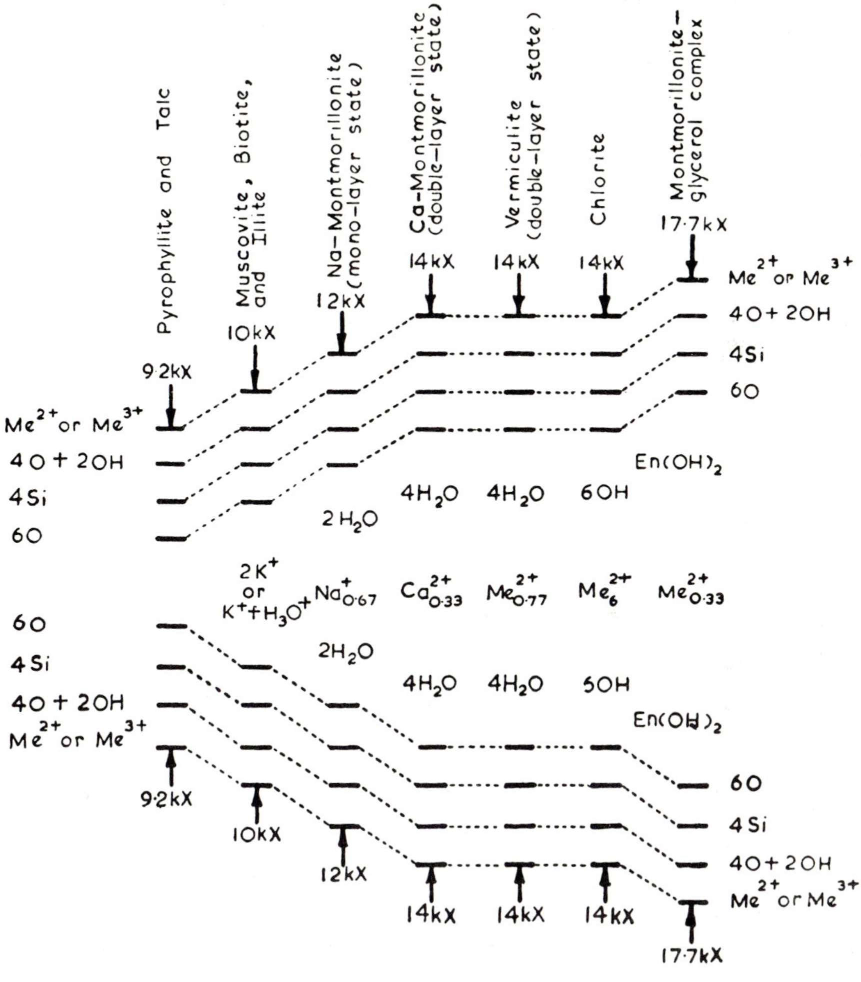

Figure 9. Possible mixed-layering of clay minerals (*after* Cole and Hosking, 1957, p. 249). Reprinted by courtesy of the Mineralogical Society, London.

Other clay minerals that are thus regularly interstratified are *corrensite* (*see* under chlorite, p. 34), and a number of minerals that have long spacings (Sudo, 1959), as well as some 1 : 1 montmorillonite-chlorite regular interstratifications (MacEwan and others, 1961, p. 441-442).

In random mixed layering the diffraction spacings may be calculated (MacEwan and others, 1961, p. 397-418), and the diffraction patterns can be identified from charts calculated for different proportions of end members producing *d* spacings, assuming that the interstratified minerals have the same structure factor, that is, the same type of structure. Diffraction patterns from randomly mixed-layer minerals do not give an integral series of *d* spacings from *d* (001).

X-Ray Identification of Mixed-Layer Clays

(1) Regular interstratified clays are identified by an integral sequence of 00*l* reflections representing the sum of the layer thicknesses; for example, two layers of 14 Å minerals will give a 28 Å spacing for (001). It may be necessary to apply various treatments to identify the components.

(2) Randomly interstratified clays are more difficult to interpret. However, inspection of the diffraction chart may suggest identification but it should be remembered that the sequence of reflections from a mixed-layer clay is short. Random interstratification of two or three different clays is very common. In a preliminary analysis of a two-component system, no large errors result if it is assumed that the diffraction peaks move linearly from the position for one pure component toward the nearest position of the other component as the proportion of the components alter. If the distances from neighboring A and B positions are x and y, respectively, the proportion of A is deduced to be y/(x+y) (MacEwan and others, 1961, p. 432-433). Mixtures of three components have been studied by Weaver (1956). Jonas and Brown (1959) used an empirical method to estimate the first-order spacing from three-component interstratified systems. Their determinative diagram is reproduced as Figure 10.

In a diffraction pattern, mixed layering of mica is indicated by a series of reflections on the low-angle side of the 10Å *d* spacing. The resultant spacing is broad and has several peaks making up the "hump" or shoulder. If water is present it may be removed with heat, and the structure will revert to that of a mica; if the mica is interlayered with organic matter, the latter may be removed. Saturation with K^+ ions may reconstitute the mica structure. Mixed layering of montmorillonite, vermiculite, and chlorite is indicated by reflections on the high-angle side of the 14 Å *d* spacing. Glycolation and heating are generally required to identify the components that form the shoulder or hump on the high-angle side of the 14 Å spacing.

The recognition of regular or random interstratification of clay minerals shown in a diffraction pattern is the first step in the identification of the components present. After the identification of these components by various treatments and the preparation of diffraction patterns from the treated materials, the investigator will generally be able to recognize the type of interstratification given by certain components.

Mechanical mixtures of clay minerals can be readily identified as the *d* spacings of prominent reflection planes; all appear in a diffractogram, and each reflection plane can be identified as belonging to a particular mineral. The amounts of the components in a mixture that can be recognized in a diffractogram are discussed under "Quantitative estimates of clay minerals in mixtures."

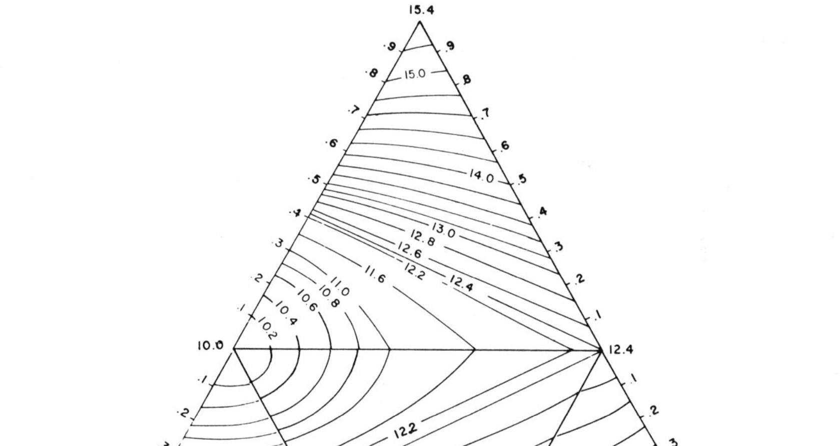

Figure 10. Three-component composition triangles for the four possible interlayer mixtures of 10 Å, 12.4 Å, and 15.4 Å clays. The apparent X-ray diffraction spacing, *d*, is contoured (*after* Jonas and Brown, 1959, p. 84). Reprinted by courtesy of Dr. E. C. Jonas, the Journal of Sedimentary Petrology, and Society of Economic Paleontologists and Mineralogists.

PALYGORSKITE-SEPIOLITE GROUP MINERALS

The crystalline structure of the minerals in this group is similar to that of the amphiboles; it consists of long double chains of composition Si_4O_{11} which are parallel to the length of the mineral fibers. Between these double chains there are channels occupied by water (Bradley, 1940). The channel water is removed stepwise by heating, but all the water is not removed until about 850°C is reached. As water is removed, the structure remains rigid until 350°C-400°C, when contraction occurs; the original structure is completely destroyed at 850°C-1000°C (Caillère and Hénin, 1961a; 1961b, p. 337-338, 350). There is no expansion with organic liquids.

Early descriptions of the Palygorskite-Sepiolite Group suggested an isomorphous series between the Al end member (palygorskite) and the Mg end member (sepiolite), but more detailed investigations showed that this is not so, and that there are two separate minerals with slightly different structure. In the palygorskites there is isomorphous replacement of Al by Mg, but in the sepiolites there is no such replacement (Millot, 1964, p. 36).

Palygorskite

Palygorskite, pilolite, lassalite, and attapulgite are synonymous; the name palygorskite has precedence. Attapulgite, the clay from Attapulgus, Georgia, is the name used for palygorskite in the United States.

Chemical Composition. "The ideal formula for the half-unit cell is $Si_8Mg_5O_{20}$ $(OH)_2$ $(OH_2)_4$ $4H_2O$, where OH_2 denotes bound water and H_2O represents water held in channels." (Caillère and Hénin, 1961b, p. 346.) Isomorphous replacements occur with Fe and Ca; the Ca present in many analyses is an exchangeable ion (Bradley, 1940).

Unit Cell. a = 12.9 Å; b = 18.0 Å; c = 5.3 Å (Caillère and Hénin, 1961b, p. 346).

Diffraction Pattern. As the structure of palygorskite is similar to that of sepiolite, the same general features are found in the diffraction pattern. Reflections of sufficient intensity to appear on diffractograms occur at 10.5, 4.5, 3.23, and 2.62 Å.

Identification. Palygorskite is identified from its diffraction pattern. Above 400°C the 10.5 Å reflection is replaced by a broad band, and near 800°C the structure is completely destroyed. As palygorskite and sepiolite do not have identical crystal forms or cell dimensions, the d spacings differ for the two minerals. For example, *d* (110) is 10.48 Å for palygorskite and 12.07 Å for sepiolite; *d* (130) is 5.44 for palygorskite and 7.48 Å for sepiolite; *d* (200) is 6.45 Å for palygorskite and 6.73 Å for sepiolite (Calliére and Hènin, 1961a, p. 340; 1961b, p. 352).

Occurrence. Palygorskite has been recorded as a hydrothermal alteration product of pyroxenes and amphiboles but the largest deposits are of authigenic origin, as in the interbedded marine to fresh-water lagoonal deposits in the Hawthorn Formation (Miocene) of Georgia and Florida (Kerr and others, 1950, p. 45). Palygorskite is more abundant than sepiolite, and numerous occurrences of it have been cited in the literature. Like sepiolite, palygorskite requires an alkaline chemical environment for survival. It has been identified in alkaline unleached soils from many localities.

Sepiolite

Sepiolite is known as the clay meerschaum, a lightweight white or gray clay. Meerschaum = sea-froth (German) and sepiolite = cuttlefish (Greek).

Chemical Composition. A hydrous magnesium silicate the structural scheme of which provides sites for 12 tetrahedral ions, 9 octahedral ions, 40 articulated oxygens and hydroxyls and about 6 molecules of low temperature water per half unit cell (Nagy and Bradley, 1955, p. 888). The ideal formula can be rationalized to:

$$H_6Mg_8Si_{12}O_{30}\ (OH)_{10} \cdot 6H_2O$$

The formula calculated from analysis of sepiolite from Little Cottonwood, Utah, by Callière and Hénin (1961, p. 333) is:

$$(Si_{11.67}Al_{0.24}Fe^{+3}{}_{0.09})\ (Fe^{+3}{}_{0.02}Mn^{+2}{}_{0.53}Mg^{+2}{}_{7.42})\ O_{32}\ Cu_{0.15}$$

Other analyzed sepiolites give similar formulas.

Unit Cell. a = 13.50 Å; b = 26.97 Å; c = 5.26 Å (Callière and Hénin, 1961a, p. 335).

Diffraction Pattern. Sepiolite has a wide range of crystallinity. A diffractogram of well-crystallized sepiolite shows a number of *d* reflections between 7° and 44° 2θ (12.07 Å to 2.07 Å) (Callière and Hénin, 1961a, p. 336). Poorly crystalline sepiolite shows only the strongest *d* reflections in a diffractogram; these are at about 12.6 Å, 4.31 Å, and 2.62 Å. Callière and Hénin (1961a, p. 336) illustrate the diffractograms given by well-crystallized and poorly crystalline sepiolite.

Identification. By *d* spacings on a diffractogram and by characteristic dehydration; loss of zeolitic water $< 250°C$, bound water

250°C-620°C, and of hydroxyls 620°-1000°C (Callière and Hénin, 1961a, p. 334). The structure is not modified by loss of water < 300°C, but irreversible dehydration occurs at about 250°C. Above 350°C the structure is changed. The 12 Å reflection becomes weak and diffuse, a new *d* spacing appears at 9.8 Å, and the intensity of the spacing at 7.6 Å increases. This structure is stable < 700°C, but on further heating the mineral becomes amorphous to X-rays (Callière and Hénin, 1961a, p. 337-338).

Occurrence. Sepiolite has two principal modes of occurrence in both of which stability is maintained only under alkaline conditions. Sepiolite is decomposed by acid leaching, and therefore not to be expected in situations in which the pH is less than 7. One type of occurrence is in vein alteration of original biotite, as in the sequence biotite → vermiculite → sepiolite. Sepiolite also occurs as crack fillings in altered serpentine. The second type of occurrence is in soils with underlying dolomitic marl, often of lacustrine origin. Recently sepiolite has been recorded from the marine sediments of the Mid-Atlantic Ridge by Hathaway and Sachs (1965) and by Siever and Kastner (1967). Sepiolite has formed there authigenically as an alteration product of serpentine by marine weathering. Millot (1964, p. 233-234) has summarized the authigenic occurrences of sepiolite.

X-Ray Diffraction as Applied to Clay Mineralogy

It is presumed that the majority of readers are familiar with the use of an X-ray unit. Details of the theory of X-ray diffraction are given in Klug and Alexander (1962) and of the operation of the various types of X-ray units in descriptive instructions supplied by the manufacturers. The principles are the same for all units, but there are slight differences in the details of operation.

X-ray diffraction using a wide-range goniometer results in pulse generation by the sample being recorded on a chart. This method is peculiarly suited to the examination of fine-grained samples. In fact, fine grinding is desirable for the X-ray diffraction of all samples except thin slabs of rock; clay should be separated from other minerals by dispersion and sedimentation. A chemical analysis gives only the amount of chemical elements that are present; it does not give information about how these elements are combined with one another. X-ray diffraction apparatus gives a pattern of the principal crystallographic planes that cause the diffraction of X-rays, and hence, by interpretation, gives the actual minerals that are present in a mixture. As the mass absorption coefficient (μ^*) does not differ greatly for the majority of the minerals to be identified in clays, the orientation of the grains is of considerable importance. However, a direct comparison of the amounts of a mineral in a mixture as presented by peak intensities or areas is not possible (*see* Quantitative Estimation of Clay Minerals in Mixtures, p. 63.)

The clay minerals as described above all have basically similar layer-lattice structures except the Palygorskite-Sepiolite Group (Table 1). Thus, all exhibit similar X-ray diffraction features, but as the unit cells and chemical character vary in different minerals, the identification of the individual minerals requires tests other than diffraction to be applied for their identification. The X-ray diffraction patterns clay minerals give, in oriented mounts, the basal spacings ($00l$) of the minerals; in unoriented mounts the *hkl* spacings are given. The basal spacing gives periodicities normal to the plane along c^*; this dimension may vary with humidity, dehydration, glycolation, cations in interlayer positions, acid treatment, etc. Identification of clay minerals depends on the experimental reaction of the individual mineral under controlled conditions. The *hk* spacings are the *a* and *b* directions of the unit cell in all minerals except those of the triclinic class. These spacings do not vary appreciably with treatment.

Two important considerations in X-ray diffraction are the correct alignment of the X-ray instrument and of the sample in the path of the X-ray beam so that maximum diffraction is obtained. The mass absorption coefficient (μ^*) varies with the chemical composition of the sample and with the radiation used. However, for the minerals commonly found in clays it ranges between 30 and 50 except for the Fe-chlorites. It is generally less with $CuK\alpha$ than with $FeK\alpha$. The mass absorption coefficients for a number of common minerals are given in Table 7.

As the mass absorption coefficient of iron is greater with $CuK\alpha$ than with FeK_α radiation, the latter is preferred for minerals known to contain a high percentage of iron (compare μ^* for Mg- and Fe-chlorites in Table 7). With the curved crystal monochrometers that are available and in use in some laboratories, the optimum radiation for the identification of clay minerals in samples that contain an appreciable amount of iron can be easily found.

One of the requirements for the efficient use of X-ray diffractometry in identification of minerals is the correct thickness of the mounted sample to give the maximum intensity for diffraction from crystal lattice planes (*see* p. 48). Another is the requirement of size of grains to be X-rayed. The grains should be size-graded as they are in the preparation of a clay fraction. The intensity of X-ray diffraction, obtained with different sizes of less than 325-mesh quartz powders using $CuK\alpha$ radiation, is given in Table 8.

It is apparent that the clay fraction < 2 microns gives the best results for X-ray diffraction and that the arbitrary separation of clay from silt and larger mineral grains is theoretically sound.

TABLE 7. MASS ABSORPTION COEFFICIENTS FOR CuKα AND FeKα RADIATION OF COMMON MINERALS*

Mineral	Mass absorption coefficient CuKα	FeKα
Quartz	35	67
Orthoclase	48	101
Albite	34	78
Anorthite ($An_{10}Ab_0$)	52	98
Calcite	71	140
Dolomite	50	93
Siderite	162	47
Pyrite	200	117
Kaolinite	30	60
Metakaolin	32	64
Muscovite	44	85
Illite[a]	55	73
Glauconite[b]	117	75
K-bentonite[c]	52	74
Mixed-layer[d]	59	72
Biotite[e]	94	87
Chlorite (Mg)[f]	54	57
Chlorite (intermediate)[g]	104	61
Chlorite (Fe)[h]	133	62
Montmorillonite[i]	42	66
Gibbsite	24	48
Mullite	32	64
Iron oxide, Fe_2O_3	231	58
Magnetite	238	60
Ilmenite	189	135

[a]average chemical composition of 8 type samples (Hower and Mowatt, 1966).
[b]average chemical composition of 40 glauconite samples (Hendricks and Ross, 1941).
[c]average chemical composition of 9 type samples (Hower and Mowatt, 1966).
[d]average chemical composition of 4 type samples (Hower and Mowatt, 1966).
[e]average chemical composition of 137 biotites (Foster, 1960).
[f]average chemical composition of 78 Mg-chlorites (Foster, 1962).
[g]average chemical composition of 56 (Mg,Fe) chlorites (Foster, 1962).
[h]average chemical composition of 20 Fe-chlorites (Foster, 1962).
[i]average chemical composition of 54 members of the montmorillonite-beidellite group (Ross and Hendricks, 1945).

*Calculated by Dorothy Carroll.

TABLE 8. INTENSITY OF X-RAY DIFFRACTION FROM FINE QUARTZ POWDERS

	<5 microns	5-15 microns	5-50 microns	15-50 microns
Mean area*	11,293	11,268	9,227	8,513
Mean deviation	132	236	929	1,545
Mean percent deviation	1.2	2.1	10.1	18.2

*Measured under principal *d* spacing of lattice.
After Klug and Alexander, 1962, p. 292. (Reprinted by courtesy of John Wiley & Sons, Inc.)

TECHNIQUES FOR X-RAY IDENTIFICATION OF CLAY MINERALS

The crystalline structure of the clay minerals (described above) is such that the most important diffractions from an X-ray beam occur within the 2-37° 2θ scanning distance. An oriented mount will give the first five (00*l*) basal spacings of well-crystallized chlorite or mica and the basal spacings of the other clay minerals, except the Palygorskite-Sepiolite Group which does not have the phyllosilicate structure. Such a scan will give sufficient length of diffractogram for a preliminary identification of any clay mineral present and will also include prominent reflections of quartz, cristobalite, feldspar, carbonates, as well as such minor constituents as amphiboles, pyroxenes, and zeolites. An untreated oriented mount can give little further information. If the goniometer scans at 1° 2θ per minute and the recording chart speed is 30 inches per hour, then the preliminary identification will take a little over half an hour. Many investigators sacrifice mineral information for speed by scanning at 2° 2θ per minute or even faster. The speed of scan is decided by the requirements of the results to be obtained from the sample.

Standards. The *d* spacings obtained on the initial diffractogram are next compared with the *d* spacings of standard minerals as recorded by the American Society for Testing Materials (A.S.T.M.) on their sets of file cards. It is important to have a set of A.S.T.M. file cards, together with an Index to the Powder Diffraction File for every laboratory in which clays are identified by X-rays. In addition, in the United States, samples of A.P.I. (Kerr and Hamilton, 1949, American Petroleum Institute Project 49) standard clay minerals should be available so that a set of standard diffractograms of the various clay minerals can be made for comparative purposes.

In a preliminary examination of a diffractogram of a clay sample, the diffraction angles (2θ) are converted to d spacings in Ångstrom units by means of tables (Switzer and others, 1948; U.S. National Bureau of Standards, 1950; Fang and Bloss, 1966). Comparison of these spacings with diffractograms of standard minerals and the A.S.T.M. file cards should indicate whether the sample consists of a single mineral or of a mixture of minerals. The diffractogram of Figure 11 shows a clay that is composed of several different minerals.

From a preliminary identification, certain tests are suggested for the complete identification of the individual minerals in the diffractogram. A commonly employed scheme of identification is given in Table 9.

As shown in Table 9, the principal auxiliary treatments for the identification of clay minerals are glycolation and heating. Both treatments cause distinctive expansion or contraction of the c lattice dimension that are measurable on diffractograms. With prolonged heating at high temperatures, different for each group of minerals, all three lattice parameters are affected; the crystalline structure of the mineral breaks down, and amorphous mineral matter results. Continued high temperature may cause the mineral substance to recrystallize into other minerals, but that is another phase of mineralogy. It is not required in the identification of clay minerals.

Identification of clay minerals often requires the application of a number of auxiliary techniques, such as chemical, petrological, and infrared absorption analysis, electron microscopy, differential thermal analysis, and others (Brindley, 1961a), but a clay mineralogist soon learns by experience which auxiliary treatment is required for his samples. For example, if the samples appear to contain much organic matter, this should be removed with hydrogen peroxide prior to X-raying. Similarly, it may be necessary to remove iron oxide coatings with sodium dithionite or other similar means.

Inasmuch as many laboratories may be responsible for the identification of clay minerals in large numbers of samples, routine application of methods highlighting the main differences in clay mineral groups are applied. Auxiliary techniques can then be applied if necessary, provided the equipment and skilled personnel

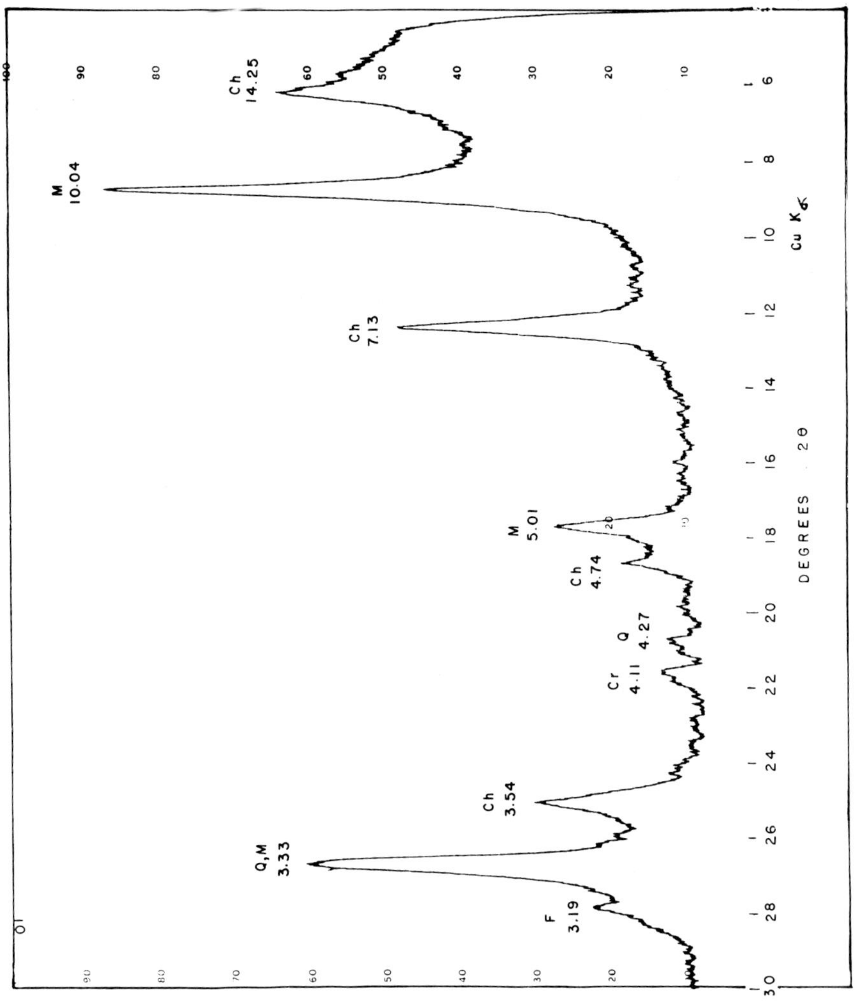

Figure 11. X-ray diffractogram of oriented mount of clay typical of sediment 50 cm below sea-floor at lat 35° 00′ N., 165° 00′ W., northern Pacific Ocean (laboratory working chart, Dorothy Carroll, unpub.). Ch, chlorite; M, mica; Q, quartz; Cr, cristobalite; F, feldspar.

are available. The preliminary identifications can then be augmented and refined. In such a way the routine techniques of Table 9 can be augmented by the X-ray determination of the *a* and *b* lattice parameters. Powder photographs can also be made to obtain all the *d* spacings of a clay mineral. However, for the identification of a large number of clay-bearing samples, the speed at which diffractograms can be made (15 or 30 minutes) has made the X-ray diffractometer the choice of most investigators. In a well-operated laboratory with sufficient competent assistance, the clay minerals in at least 20 samples can be identified in a working day of 8 hours.

Routine Procedure for the X-ray Identification of Clay Minerals

(1) Make oriented mount on tile or Vycor glass slide using slurry from the dispersion of original sample. Clay should be < 2 micron particle size.

(2) Allow clay to dry at room temperature.

(3) Place in desiccator at constant humidity when clay appears dry (4-8 hours for original drying). Some localities are too dry, others are too humid, and montmorillonites and halloysites react to humidity.

(4) Make diffractogram (A) at predetermined best setting of instrument and electronic panel to give a diffractogram that shows the *d* spacings clearly (this will require some trial and error procedures with standard samples; goniometer should be carefully checked for alignment). Do not remove chart from recorder.

(5) Identify and label *d* spacings shown on diffractogram.

(6) Place mounted sample in small desiccator containing about ½ pint ethylene glycol (Brunton, 1955). Run 6-12 samples to this stage. When desiccator has room for no more mounted samples, place in oven at 60°C for at least one hour (several hours may be required for expansion to take place in some minerals). Remove desiccator and set aside to cool; leave samples in desiccator until required.

(7) Make a diffractogram of each glycolated sample (B) on same chart with original sample (A), but with different colored ink and with zero set above the (A) zero. It is convenient to make diffractograms of about six samples (A), then wind chart back and make (B) above (A).

(8) Inspect and label *d* spacings in (B), particularly any spacing at about 14-17° 2θ.

TABLE 10. TEMPERATURES REQUIRED IN THE IDENTIFICATION OF CLAY MINERALS

Mineral	Temperature, 1 hour	Effect
Kaolinite, well-crystallized	575-625°C	Replacement by amorphous metakaolin; no diffraction pattern
Kaolinite, disordered ("fire-clay")	550-562°C	Similar to kaolinite
Dickite	665-700°C	Similar to kaolinite
Nacrite	625-680°C	Similar to kaolinite
Halloysite	125-160°C	Water removed
	560-605°C	Similar to kaolinite
Metahalloysite	125-150°C	Water removed
	560-590°C	Similar to kaolinite
Allophane	140-180°C	Adsorbed water removed; amorphous at all temperatures
Mica, well-crystallized	700°C	Shows gradual loss in weight, but does not break down below 700°C; (001) spacings remain in diffractograms below 700°C and up to 1000°C
Illite, and clay micas	125-250°C	Loss of hydroscopic water
	350-550°C	Reverts to mica structure
	700°C	Similar to mica
Glauconite	530-650°C	Loss of interlayer water; reverts to mica structure
Celadonite	500-600°C	Similar to glauconite
Biotite	700°C	Phlogopite is similar to muscovite Biotite shows breakdown 700-1000°C

(9) At this stage the preliminary identification of the principal clay minerals present in the mount will have been made (*see* descriptions of minerals).

(10) Heat in a muffle furnace at calibrated controlled temperature to give the characteristic heat reaction of the various minerals. The heat required for the identification of the common clay minerals is given in Table 10.

After heating and cooling in a desiccator, make a diffractogram

TABLE 10. (Continued)

Mineral	Temperature, 1 hour	Effect
Vermiculite	<300°C	Water removed in stages with 14Å 13.8Å 11.6Å 9Å. Initial (001) spacing controlled by humidity
Montmorillonite Group	300°C	Original 15Å spacing disappears; 9Å spacing develops
Chlorite Group	600-800°C	Show gradual weight loss, but no structural change
Mg-chlorite	650°C	14Å spacing is intensified; (004) at 3.54 not affected
Fe-chlorite	500°C	14Å spacing less intense and may become broad and diffuse
Mixed-layer clays	<600°C	Varies with kinds of minerals present and amount of each; check with X-rays (Cole and Hosking, 1957)
Palygorskite-Sepiolite Group		Water occurs in channels in structure
Sepiolite	<200°C	Rapid dehydration
	>200°C	Slow dehydration; spacing at 12Å above 350°C becomes weak and diffuse; a spacing at 9.8Å appears and spacing at 7.6Å becomes more intense; recrystallizes at 800°C
Palygorskite	400-440°C	Rapid dehydration, but no change in structure <400°C; >400°C 10.5Å spacing becomes broad and diffuse; near 800°C structure is destroyed

Compiled by Dorothy Carroll

(C) above (A) and (B) on the same chart with a different color and zero setting.

Blue (A), green (B), and red (C) inks on a recorder chart make the results of the various treatments very easily comparable, and it is advantageous to have all three results recorded on the same chart. Interpretation of the diffractograms and filing of results is thus simplified.

As the X-ray identification of clay minerals is generally on samples < 2 microns in grain diameter, the effect of crystallite size on the diffraction of the *d* spacing need not be considered in the majority of determinations; the size of the clay fraction particles is a constant factor in the identification. However, one of the uses to which diffractograms can be put is the determination of crystallite size. If the crystallites are extremely small, a broadening of the *d* spacings is found, as shown by Klug and Alexander (1962, p. 530-538). Such line broadening may be hard to detect, because the mineral may be disordered and thus have line broadening because of crystallinity rather than crystallite size. In some clay fractions separated from sedimentary materials, a line broadening in quartz indicates that the quartz grains are of extremely fine sizes.

It has been shown by Keely (1966) that the surface area of mineral grains of different sizes can be measured by an X-ray line broadening effect. Keely's investigation stems from earlier investigations of the relation between crystallite size and approximate specific surface of metallic oxides in their use as catalysts and coating agents. The line broadening was measured from the principal *d* spacing of the material investigated as shown on a diffractometer chart, and the mean particle size was determined from the Scherrer equation:

$$D = \frac{0.9\ \lambda}{(\beta - \beta_o)\ \cos \theta}$$

Where λ = wave-length of X-ray,

β = breadth at half-maximum intensity of a peak of the material,

β_o = breadth at half-maximum intensity of the completely crystalline material,

θ = Bragg angle (one-half the spectrometer angle).

Both D and λ are in Ångstrom units.

It is assumed that all the particles are spherical, and it is necessary to obtain the density and the volume of 1 gram of the material. The number of particles is determined by:

$$\text{Number of particles} = \frac{\text{volume of 1 gram}}{4/3\pi\ (\tfrac{1}{2} \times \text{mean dimension})^3}$$

The surface area is calculated by multiplying the number of particles by the surface ($4\pi r^2$) of each particle (r = radius).

However, clay minerals are prismatic, and in a preliminary experiment to test the application of Keely's method for the determination of surface areas of clay minerals, the equation was modified to accommodate prismatic grains by substituting the surface areas of prismatic grains using the unit cell dimensions for the term $4/3\pi$. Preliminary results by this method calculated from X-ray data for the surface area and number of particles in 1 gram of standard minerals are given in Table 11. The data have not been checked with the standard B.E.T. nitrogen absorption method.

TABLE 11. SURFACE AREA AND NUMBER OF PARTICLES IN 1 G OF SOME STANDARD MINERALS*

Mineral	Particle size of prepared mineral† micron	Surface area of 1 g, m^2	Number of particles in 1 g
Quartz	<2	7.6	5.23×10^8
Feldspar	<2	40.3	1.00×10^8
Mica ($2M_1$)	<2	11.4	129.4×10^7
Kaolinite	<2	6.7	27.8×10^6

*Calculated from line broadening of principal X-ray peak by Dorothy Carroll (unpub. data).

†Particle size obtained by sedimentation.

Preparation of Clays for X-Ray Examination

Sizing. In the Introduction, clay-sized material was defined for practical purposes as particles < 2 microns in diameter. Except in deposits of clay, the clay fraction is associated with larger mineral grains from which it must be separated prior to examination. In addition, the clay particles occur as aggregates both with other clay minerals and with other minerals. Inasmuch as clay is the agriculturally important ingredient of soils, agricultural chemists evolved a general method of making a size analysis or mechanical analysis of soils. This method can be applied in part to the preparation of clays for X-ray examination. There are many different procedures based essentially on that of the agricultural chemists. One such procedure used by the writer is given in the flow-sheet on the following page.

There are many variations of this basic procedure, but the aim of each is to obtain a dispersed sample of clay for X-ray examination. Each person processing clay for X-ray examination will eventually evolve his own technique according to the raw material, the equipment available, and the assistance he has for the preparation of samples.

Mounting of Sample. Two kinds of clay mineral mounts for X-ray diffractometry are generally made: (1) An oriented mount obtained most readily from a clay slurry obtained as part of the process of making a size distribution of sedimentary material. A similar oriented mount can be made from dry clay, but the mount may be unsatisfactory because of the difficulty of deflocculating clay that has been dried. Clay smears can also be made that are satisfactory. (2) Unoriented mounts of dry clay in aluminum holders.

Raw sample
Unconsolidated
100*g

Disaggregate in water

(a) in ultrasonic vibrator

(b) in mechanical shaker to which a peptizing agent (NH_4OH, NaOH, calgon, sodium hexametaphosphate, or other) has been added

Allow to stand in cylinder containing 1000 ml water for sufficient time to sediment the sand fraction (>62 microns) according to Stoke's Law

Siphon off clay and silt fractions in suspension

Remove silt from clay by centrifugation

Allow silt to dry at room temperature

Weigh and store

Remove water from clay by filter candle under vacuum until about 200 ml of slurry remains

Remove about 10 ml with eyedropper and place evenly on glass slide or semiporous tile to dry at room temperature

Place a sample of slurry in a snapcap vial for reference and storage

Place remainder of clay slurry in large evaporating dish to dry at room temperature. When dry, scrape up and store

*This flow sheet gives a general outline of procedure; 100 g is the maximum amount that is needed for processing. The amount of raw sediment to be used varies with mineralogical composition and with the requirements of the examination; several procedures may be used for the separated clay; for example, cation exchange capacity, electron microscopy, chemical analysis, X-ray diffraction.

The geometry of the wide-angle goniometer used in diffractometry requires that the sample to be X-rayed be perfectly flat and parallel to its mount (curved samples give more precise X-ray reflection, but they are not commonly used and present difficulties in sample mounting; flat mounts give sufficiently accurate results for the routine identification of clay minerals).

The method of mounting the fine-grained mineral to be X-rayed is of fundamental importance. Klug and Alexander (1962, p. 252) state that if the maximum diffraction intensity is required, it is necessary for the thickness of the sample, t, to be great enough to satisfy the following equation:

$$t \geqq \frac{3.2}{\mu} \frac{\rho}{\rho'} \sin\theta$$

where ρ = density of powdered mineral
ρ' = density of powdered mineral + interstices
μ = linear absorption coefficient of the mineral

Aqueous suspensions containing only a very small amount of dispersed clay may make too thin an oriented mount to cause X-ray diffraction of the clay minerals (Klug and Alexander, 1962, p. 252). If such a mount is made on a glass slide, only the X-radiation of amorphous material is produced; generally, the optimum conditions are easily produced. If the powder spread on the tile is too thin, the principal *d* spacings are weak and diffuse, and the *d* spacings of the tile material occur, depending on the composition of the tile. If the mount is too thick, a high background is produced in a diffractogram by X-ray scattering. It is advisable always to make a diffractogram of a tile if one is used as a mount. The thickness of the sample should be from 0.07-0.12 mm (Sudo and others, 1961), and this is obtained if approximately 40 mg of sample is spread over an area of 2.0 × 2.7 cm (540 mm^2). The microscope glass slides used for mounting in the United States are 4.5 × 2.6 cm (1170 mm^2), so that a comparative thickness is given by a 90 mg sample.

Oriented Mounts. There are a number of advantages in using a completely dispersed clay (as in a slurry that is produced in a mechanical analysis; *see* the flow sheet given under Sizing, p. 57) for making oriented mounts. Although the amount of clay in a slurry can be measured, there may be physical and mineralogical differences from sample to sample that make such a procedure time-consuming and impracticable. Judgment based on experience gives a sufficiently accurate amount of clay on the mount, and furthermore, the actual thickness of the dried clay can be accurately measured with a microscope as a check if necessary. There are three types of mounts that are commonly used for oriented clay diffractometry: glass slides for evaporation mounts or smears (heat resistant or Vycor glass is preferable), semiporous ceramic tile, or porous ceramic tile. The kind of mount used depends on availability (many tiles have to be cut to size), ease of manipulation during preparation and use, experience of the preparator, amount of clay in the slurry used (a clay fraction that contains much colloidal organic matter gives a very meager mount for clay minerals and a glass slide is advisable to reduce the background and eliminate spacings due to minerals in the tile), and kind of results required by the investigator. Some of the procedures used in making oriented clay mounts have been discussed by Gibbs (1965, p. 741-751), but no firm recommendations can be made that will apply to every clay that is examined by X-ray diffraction. The writer prefers, after considerable experience, to make mounts on semiporous tile (Coors porcelain, type used for drying crystal precipitates, unglazed, 48 × 48 × 6 mm, cut in half, giving plates 48 × 24 × 6 mm, which fit easily into the holder of a Norelco wide-range goniometer). The clay slurry dried easily at room temperature by a combination of drainage through the tile (especially if set on a styrofoam pad) and evaporation. The writer uses a batchwise procedure that processes six samples at a time. Some investigators prefer porous tiles that are cut in thin slabs and employ vacuum to remove water from the slurry, thus leaving the clay sample on the tile which is then allowed to dry at room temperature or heated at low temperatures. There are a number of other procedures that can be used in the preparation of oriented clay mounts; their use will depend on the type of material investigated and on the experience of the investigator.

The preparation of standard minerals for X-ray diffraction requires care. Some minerals, such as quartz and feldspar, can be ground with a pestle and mortar to obtain grains of sufficient fineness. Micas, chlorites, and vermiculites should be pulverized in a small electrically powered device, such as that which was designed for this purpose in the Mineralogical Museum of Oslo University (Neumann, 1956). The grinding of micaceous minerals is difficult and fine grinding results in the production of an amorphous or allophanelike material (Fieldes and others, 1966), because the ordered crystalline structure of these minerals is destroyed. Pulverizing, followed by sieving, produces mechanically split or shredded particles that have the properties of larger grains, and such grains do not show any of the effects of grinding.

Unoriented Mounts. A sample of clay (< 2 microns) is packed into an aluminum holder following McCreery's procedure (Klug and Alexander, 1962, p. 300-301); with good technique an unoriented mount is obtained. To increase random orientation of the clay minerals, the sample should be sieved onto the holder and cut down into the hole in the mount with a small spatula held with the thin edge of the blade at right angles to the holder. When the hole in the mount is filled, it should be pressed down lightly with the wide blade of the spatula to obtain the necessary flat surface which coincides with that of the surface of the aluminum holder. A diffraction pattern of the sample thus mounted will show the amount of random orientation by the intensity of the *hkl* reflections.

A special holder for making mounts of layer-lattice minerals to obtain the *hkl* spacings with the older type of X-ray goniometers was designed by Byström-Asklund (1966) of the Swedish Geological Survey. The clay particles are packed into this holder with their basal spacings (the natural way for platy particles to lie) parallel to the base of the holder, which is then turned on its side when mounted in a diffractometer. The *hkl* reflections are readily obtained with type of mount which is especially useful for recording the (060) spacing of clay minerals. This holder is shown in Plate 1.

Quantitative Estimation of Clay Minerals in Mixtures

On superficial examination it might seem that direct comparison of intensities or peak areas of minerals identified on a diffractogram of a mixture of minerals would give an estimate of the quantities of the individual minerals present. However, this cannot be done because of differences in mass absorption coefficients of the individual minerals (Table 7), of the orientation of the grains, thickness of the mount, weight of the clay sample X-rayed, evenness of the spread of the mixed minerals, differences in crystal perfection, polytypism, hydration, chemical composition and many other factors.

Under standardized conditions, using mixtures of minerals that are equally well-crystallized, the quantity of each component can be obtained by using an internal standard (generally fluorite) and making calibration curves from which the weight of each component present in the mixture can be found (Klug and Alexander, 1962, p. 419-438). The method of internal standards is seldom applicable to the mixtures of clay minerals that commonly occur in soils and unconsolidated sediments because it is difficult, if not impossible, to obtain standards that are of similar crystallinity, or of similar chemical composition to the minerals occurring naturally. So far, it has been seldom possible to separate individual minerals in natural mixtures in order to characterize them for use as standards. The qualitative identification of clay minerals depends on comparison with standards such as the A.P.I. standard clays described on page 48. These standards have also been used for semiquantitative estimates of the components in mixtures. Brindley (1961d) discussed the principles of quantitative estimation of clay minerals in mixtures, and earlier work in this field (with the exception of the Japanese investigations) was summarized by Carroll (1962b). In a recent review of clay mineralogy, Mackenzie and Mitchell (1966, p. 62) conclude that "Most clay mineralogical analyses must be regarded at best . . . as semi-quantitative."

Van der Marel (1966) published a method of making quantitative estimates of clay minerals in soils and other natural materials using X-ray, thermo- and infrared analyses. He pointed out the difficulties inherent in the composition of clay minerals and in mixtures, no part of which can be satisfactorily standardized except quartz, calcite, and dolomite, which are diluents of clay. Earlier, quantitative investigations in this field were those by Willis and others (1948). This paper tabulated the composition of 53 mixtures of standard minerals (based on the weathering sequence of clay minerals) and compared the intensity of X-ray reflections of clay fractions with those of the standards. No details of the standards were given, but the paper indicates the complexity of procedures necessary for quantitative analysis.

One of the most comprehensive quantitative investigations of soil clays was made by Norrish and Taylor (1962), who used samples of minerals < 5 microns, and either CuKα or CoKα radiation. Their method depends on the intensity of diffraction and on the mass absorption coefficient of the mineral to the radiation used. Certain minerals in soil clays can be directly estimated without the use of internal standards or calibration curves. It is particularly useful for estimating the iron contents of samples.

Quantitative estimates of shale composition using X-ray diffraction were described by Schultz (1964). He used finely ground shale powder. From diffractograms and an evaluation of the results of chemical analyses of the shale, together with heat treatments and glycolation of the samples, he was able to make quantitative estimates of the composition of the mineralogical composition from X-ray diffractograms.

Some of the problems of making rapid clay mineralogical analyses were described by Oinuma and Kobayashi (1961) and were elaborated by Sudo and others (1961), who gave details of the procedures involved in making such analyses. The methods described have been used in a number of Japanese descriptions of the mineralogy of sediments. The latter paper is particularly valuable for the practical procedures which must be considered in the successful quantitative estimation of the amounts of the individual groups of clay minerals in a mixture. One very important feature is the size of the clay sample to be used for X-ray examination and the even spread of this sample over the area to be irradiated. As Norrish and Taylor (1962) did, Sudo and others (1961) used oriented mounts of the clay minerals, both known and in mixtures. By experiment, Oinuma and Kobayashi (1961) found that a sample weighing 40 mg, spread over an area of 2.0×2.7 cm (540 mm^2) on a glass slide, gave a thickness of 0.07-0.12 mm which gave maximum intensity of diffraction of X-rays.

A method for the quantitative estimation of clay- and silt-sized minerals in mixtures was recently investigated by Carroll (unpub. data), who used weight-percent mixtures of standard minerals and took into consideration the mass absorption coefficient and the efficiency of the Geiger counter on the X-ray unit used. The results indicate that this method will be useful for quantitative estimates within the tolerance of the method and equipment. By comparison of the peak areas of the principal spacing of certain minerals, an estimate can be made of the quantity of each present. For example, kaolinite and disordered kaolinite; 2 M mica and illite, 1 Md. Measurement of the areas of the principal (001) spacings on diffractograms of minerals with a polar planimeter affords a useful method of comparing quantities of a mineral in a mixture, provided that the mass absorption and the crystallinity are similar. Figure 12 shows the results of comparing the measured areas of the (001) spacing of kaolinite with the (001) spacing of montmorillonite.

Previously, Hathaway and Carroll (1954), following the investigations of Talvenheimo and White (1952) had shown that the peak area ratio of the (001) spacing of montmorillonite to the (001) spacing of kaolinite was about 5 : 1 in mixtures of the two minerals. Figure 12 also shows this areal relation.

In quantitative estimates of clay mineral components, the writer adheres to the same techniques in the preparation of the samples, the mounting, and the method of obtaining X-ray diffractograms. A set of diffractograms of standard minerals mixed in weight-percent amounts traced on plastic sheets are used as templates for comparing X-ray spacings of minerals in mixtures with standards.

Quantitative estimates of the amounts of the clay minerals have been stated by various investigators to be probably accurate to about 5-10 percent. This is certainly true for quartz, feldspar, calcite, dolomite, 2 M mica, ferric oxide, and halite. Table 12 shows

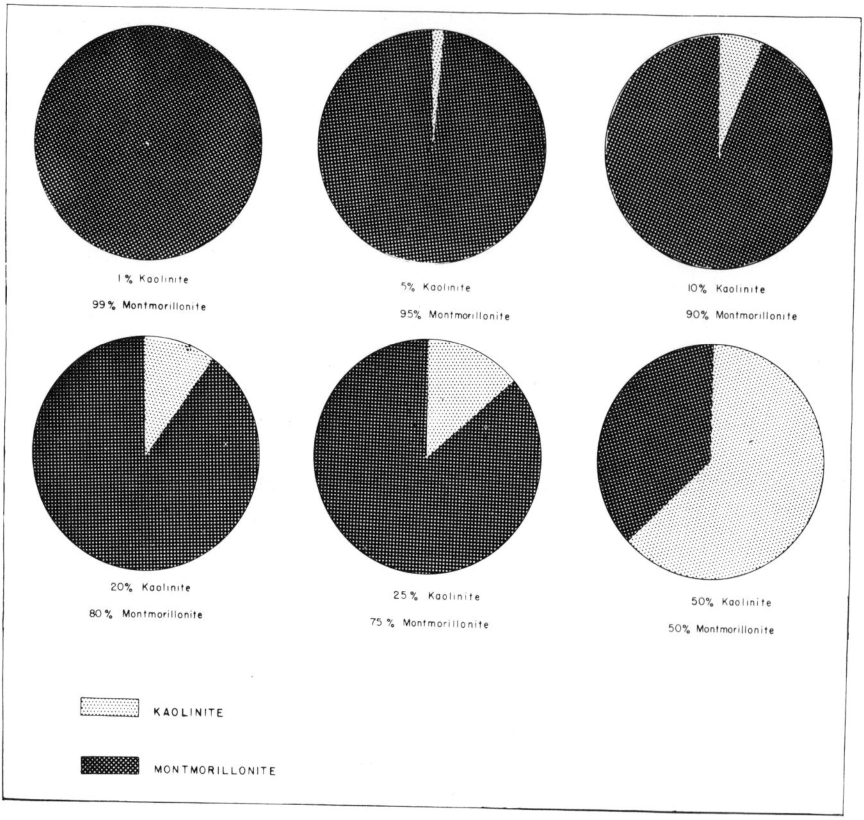

Figure 12. Illustration of the areal percentage relation of kaolinite (001) and montmorillonite (001) (Dorothy Carroll, unpub.).

the lowest percentage of clay minerals and others commonly encountered that have been reported in the literature.

Earlier investigators estimated the relative proportions of clay minerals identified in X-ray diffractograms of the clay fraction of sedimentary materials (including soils). Such an estimation, which makes use of the mass-absorption coefficient of standard minerals and indirectly of their crystallinity in evaluating the area of the principal X-ray diffraction (spacing) of the minerals, would seem to be precise enough for the methods generally used to obtain the samples in the field. Statistical sampling, not often applicable, would add refinement to the method, as would knowledge of the approximate percentage of one clay mineral in a mixture.

TABLE 12. QUANTITATIVE IDENTIFICATION BY X-RAY DIFFRACTION OF CLAY MINERALS AND CERTAIN OTHERS IN THE CLAY FRACTION (<2 MICRONS) OF SEDIMENTARY MATERALS (Cu $K\alpha$; kv 50, ma 20, 1° SCATTER SLIT, 006″ RECEIVING SLIT, CPS. 10^3, MULTIPLIER 1, RATEMETER 4; ORIENTED MOUNT ON TILE) 1° 2θ PER MINUTE; CHART SPEED 30″ PER HOUR

Mineral	Lowest percentage identifiable	*hkl*	Principal *d* spacing used for identification, Å	2θ°, (approx.)
Kaolinite	5	001	7.13-7.16	12.36-12.4
Kaolinite, disordered	1-5	001 (broad)	7.15	12.38
Halloysite, $4H_2O$	10	001	10.1	8.75
Halloysite, $2H_2O$	10	001	7.21	12.26
Mica, $2M_1$	1	002	9.99-10.4	8.85-8.80
Illite[a]	5	001 (broad)	10.16	8.70
Glauconite[b]	?	001	10.06-10.19	8.79-8.67
Chamosite[c], Fe^{+2}	?	001	7.10	12.46
Chamosite[c], Fe^{+3}	?	001	7.04	12.56
Biotite	1	002	10.1	8.77
Vermiculite[d]	5	001	14.2	6.22
Montmorillonite Group[e]	10	001	15.4 (variable)	5.7
Chlorite, Mg	1	001	14.1-14.2	6.27-6.22
Chlorite, Fe	10	001	14.1-14.2	6.27-6.22
Mixed-layer clays[f]				
Palygorskite	5	110	10.2-10.48	8.66-8.46
Sepiolite	?5	110	12.05-13.3	7.33-7.18
Quartz	1	101	3.34	26.66
Quartz	5	100	4.26	30.85
Cristobalite, α	5	101	4.04	22.0
Feldspar, orthoclase	5	040,202	3.25-3.28	27.4-27.16
Feldspar, plagioclase	5	002,040	3.17-3.18	28.14-28.0
Calcite	5	104	3.03	29.47
Dolomite	1	104	2.88-2.89	30.05
Ferric oxide[g]	<1			
Halite[h]	5	200	2.83-2.84	31.72-31.61

[a]mixed-layer 1b mica polytype.
[b]varies from mixed-layer type (authigenic) to recrystallized mica-type (consolidated sediments).
[c]poorly crystalline and may be oxidized to disordered kaolin-like mineral; not identifiable by X-ray diffraction alone.
[d]may be interstratified with biotite; heating and glycolation (negative) necessary for identification.
[e](001) spacing dependent on hydration and exchangeable cation present.
[f]*d* spacings variable due to nature of minerals interstratified.
[g]finely crystalline or amorphous coatings on mineral particles.
[h]dry clay not dispersed in water.

Compiled by Dorothy Carroll

Other useful X-ray techniques for the quantitative estimation of minerals that can be applied to clay mineral samples are the ratio of calcite to dolomite in mixtures (Gulbrandsen, 1960a, p. 93; 1960b), and the modal analysis of felsic rocks (Tatlock, 1966). Refinement of X-ray interpretative techniques in a method of chlorite identification by Brown (1955), led Schoen (1962) to investigate the possibilities of a semiquantitative analysis of chlorites by X-ray diffraction. In this method, the Mg-chlorites can be separated from the Fe-chlorites. Shirozu (1958) discussed X-ray and optical methods for the determination of the composition of well-crystallized chlorite using FeKα radiation. Previously, Brindley and Gillery (1956) discussed variation in *d* (001) with replacement of Si by Al in the tetrahedral layer of chlorites. Brindley (1961c) summarized many of the recent investigations of chlorite composition by X-ray diffraction. While it is possible to determine the chemical composition of well-crystallized chlorite quantitatively, most of the chlorite in soils and sediments is poorly crystalline and gives very poor diffractograms. So far, no method has been found either to determine its chemical composition or quantity, although rigorous examination along the above-mentioned lines will probably enable this to be done. Because of the more definite characteristic properties, these difficulties do not occur with other minerals, with the exception of montmorillonites. The recognition of polytypes in micas has eliminated some of the difficulties in the quantitative estimation of illite and detrital muscovite. At present, the two principal chlorite polytypes, 11b and 1b, are recognizable by X-ray diffraction and it is possible to estimate the approximate of each in a mixture. By application of Schoen's (1962) methods of chlorite determination, a semiquantitative analysis of the chlorite composition of a clay can be made (Carroll, 1969).

References Cited

Bailey, S. W., 1963, Polymorphism of the kaolin minerals: Am. Mineralogist, v. 48, no. 11-12, p. 1196-1209.

Barshad, I., 1948, Vermiculite and its relation to biotite as revealed by base-exchange reactions, X-ray analyses, differential curves, and water content: Am. Mineralogist, v. 33, no. 11-12, p. 655-678.

Bates, T. F., Hildebrand, F. A., and Swineford, A., 1950, Morphology and structure of endellite and halloysite: Am. Mineralogist, v. 35, no. 7-8, p. 463-484.

Bates, T. F., and Hinckley, D. N., 1959, Mineralogy and petrology of the kaolin clays of the Piedmont and Coastal Plain regions of southeastern United States: Pennsylvania State Univ. Mineral Industries Expt. Sta. Ann. Prog. Rept., June 1, 1958 to June 1, 1959, 41 p.

Bayliss, P., Loughnan, F. C., and Standard, J. C., 1965, Dickite in the Hawkesbury Sandstone of the Sydney Basin: Am. Mineralogist, v. 50, no. 3-4, p. 418-426.

Berry, R. W., and Johns, W. D., 1966, Mineralogy of the clay-sized fractions of some North Atlantic-Arctic Ocean bottom sediments: Geol. Soc. America Bull., v. 77, no. 2, p. 183-195.

Bradley, W. F., 1940, The structural scheme of attapulgite: Am. Mineralogist, v. 25, no. 6, p. 405-410.

—— 1950, The alternating layer sequence of rectorite: Am. Mineralogist, v. 35, no. 7-8, p. 590-595.

Brindley, G. W., 1961a, Experimental methods, *in* G. Brown, *Editor,* The X-ray identification and crystal structure of clay minerals: 2d ed., Mineralog. Soc. London, p. 1-50.

—— 1961b, Kaolin, serpentine, and kindred minerals, *in* G. Brown, *Editor,* The X-ray identification and crystal structure of clay minerals: 2d ed., Mineralog. Soc. London, p. 51-131.

—— 1961c, Chlorite minerals, *in* G. Brown, *Editor,* The X-ray identification and crystal structure of clay minerals: 2d ed., Mineralog. Soc. London, p. 242-296.

—— 1961d, Quantitative analyses of clay mixtures, *in* G. Brown, *Editor,* The X-ray identification and crystal structure of clay minerals: 2d ed., Mineralog. Soc. London, p. 489-516.

Brindley, G. W., 1966, Discussions and recommendations concerning the nomenclature of clay minerals and related phyllosilicates, *in* S. W. Bailey, *Editor,* Clays and clay minerals: 14th Natl. Conf. Proc., Pergamon Press, New York, p. 27-34.

Brindley, G. W., Bailey, S. W., Faust, G. T., Forman, S. A., and Rich, C. I., 1968, Report of the Nomenclature Committee (1966-67) of the Clay Minerals Society: Clays and Clay Minerals, v. 16, p. 322-324.

Brindley, G. W., and Gillery, F. H., 1956, X-ray identification of chlorite species: Am. Mineralogist, v. 41, no. 3-4, p. 169-186.

Brindley, G. W., and MacEwan, D. M. C., 1953, Structural aspects of the mineralogy of clays, *in* Ceramics, A Symposium: British Ceramic Soc., Stoke-on-Trent, p. 15-59.

Brindley, G. W., and Youell, R. F., 1953, Ferrous chamosite and ferric chamosite: Mineralog. Mag., v. 30, no. 220, p. 57-70.

Brown, B. E., and Bailey, S. W., 1962, Chlorite polytypism, Pt. 1, Regular and semi-random one-layer structures: Am. Mineralogist, v. 47, no. 7-8, p. 819-850.

Brown, G., 1955, The effect of isomorphous substitutions on the intensities of (002) reflections of mica- and chlorite-type structures: Mineralog. Mag., no. 30, p. 657-665.

—— 1961, The X-ray identification and crystal structure of clay minerals: 2d ed., Mineralog. Soc. London, 544 p.

Brown, G., and Weir, A. H., 1963, The identity of rectorite and allevardite, *in* Internat. Clay Conf. Proc., Stockholm, Sweden, Pergamon Press, London (1965), v. 1, p. 27-37.

Brunton, G., 1955, Vapor pressure glycolation of oriented clay minerals: Am. Mineralogist, v. 40, no. 1-2, p. 124-126.

Brydon, J. E., Clark, S. J., and Osborne, V., 1961, Dioctahedral chlorite: Canadian Mineralogist, v. 6, no. 5, p. 595-609.

Bubenicek, L., 1960, Developpement diagenetique des chlorites de la minette lorraine: Acad. Sci. Comptes Rendus, no. 251 (5), p. 765-767.

Burst, J. F., 1958, Mineral heterogeneity in glauconite pellets: Am. Mineralogist, v. 43, p. 481-497.

Byström-Asklund, A. M., 1966, Sample cups and a technique for sideward packing of X-ray diffractometer specimens: Am. Mineralogist, v. 51, no. 7, p. 1233-1237.

Calliere, S., and Hénin, S., 1961a, Sepiolite, *in* G. Brown, *Editor,* The X-ray identification and crystal structure of clay minerals: 2d ed., Mineralog. Soc. London, p. 325-342.

—— 1961b, Palygorskite, *in* G. Brown, *Editor,* The X-ray identification and crystal structure of clay minerals: 2d ed., Mineralog. Soc. London, p. 343-353.

Carroll, D., 1958, Role of clay minerals in the transportation of iron: Geochim. et Cosmochim. Acta, v. 14, no. 1-2, p. 1-27.

Carroll, D., 1959, Ion exchange in clays and other minerals: Geol. Soc. America Bull., v. 70, no. 6, p. 749-780.

—— 1962a, Rainwater as a chemical agent of geologic processes—a review: U.S. Geol. Survey Water-Supply Paper, 1535-G, 18 p.

—— 1962b, The clay minerals *in* H. B. Milner, *Editor*, Sedimentary petrography: Macmillan, New York and London, v. 2, p. 288-371.

—— 1969, Chlorite in central North Pacific Ocean sediment: Internat. Clay Conf. Proc., Tokyo, v. 1, p. 335-338.

Carroll, D., and Hathaway, J. C., 1963, Mineralogy of selected soils on Guam, with a section on description of soil profiles by C. H. Stensland: U.S. Geol. Survey Prof. Paper 403-F, 53 p.

Cole, W. F., and Hosking, J. S., 1957, Clay mineral mixtures and interstratified minerals, *in* Mackenzie, R. C., *Editor*, The differential thermal investigation of clays: Mineralog. Soc. London, p. 248-274.

Eggleston, R. A., and Bailey, S. W., 1967, Structural aspects of dioctahedral chlorite: Am. Mineralogist, v. 52, no. 5-6, p. 673-689.

Fang, J. H., and Bloss, F. D., 1966, X-ray diffraction tables; determination of d, $\sin^2 \theta$, or Q for 0.01° intervals of $2\,\theta$ for wavelengths $K\alpha$, $K\alpha_1$ $K\alpha_2$, and K_β of copper, iron, molybdenum, and chromium radiations and for La_1 of tungsten: Southern Illinois Univ. Press, Carbondale, Ill.

Fieldes, M., Furkert, R. J., and Perrott, K. W., 1966, The nature of allophane in soils: New Zealand Jour. Sci., v. 9, no. 3, p. 599-629.

Foster, M. D., 1960, Interpretation of the composition of trioctahedral micas: U.S. Geol. Survey Prof. Paper 354-B, p. 11-49.

Foster, M. D., 1962, Interpretation of the composition and a classification of the chlorites: U.S. Geol. Survey Prof. Paper 414-A, 33 p.

Gibbs, R. J., 1965, Error due to segregation in quantitative clay mineral X-ray diffraction mounting techniques: Am. Mineralogist, v. 50, p. 741-751.

Griffin, J. J., and Goldberg, E. D., 1963, Clay-mineral distributions in the Pacific Ocean, *in* The sea—ideas and observations on progress in the study of seas, v. 3, The earth beneath the sea and history: Interscience Publishers, New York, p. 728-741.

Grim, R. E., 1953, Clay mineralogy: McGraw Hill, New York, 384 p.

Gruner, J. W., 1934, The structures of vermiculites and their collapse by dehydration: Am. Mineralogist, v. 19, p. 557-575.

Gulbrandsen, R. A., 1960a, Petrology of the Meade Peak phosphatic shale member of the Phosphoria Formation at Coal Canyon, Wyoming: U.S. Geol. Survey Bull. 1111-C, p. 71-146.

—— 1960b, A method of X-ray analysis for determining the ratio of calcite to dolomite in mineral mixtures: U.S. Geol. Survey Bull. 1111-D, p. 147-152.

Hathaway, J. C., and Carroll, D., 1954, Distribution of clay minerals and ion-exchange capacity in some sedimentary materials, *in* A.

Swineford and N. V. Plummer, *Editors*, Clays and clay minerals: Natl. Acad. Sci., Natl. Res. Council Pub. 327, p. 264-276.

Hathaway, J. C., and Sachs, P. L., 1965, Sepiolite and clinoptilolite from the Mid-Atlantic Ridge: Am. Mineralogist, v. 50, no. 7-8, p. 852-867.

Hendricks, S. B., 1938, The crystal structure of nacrite and the polymorphism of the kaolin minerals: Zeits. Krist., v. 100, no. 6, p. 509-518.

Hendricks, S. B., and Fry, W. H., 1930, The results of X-ray and microscopical examination of soil colloids: Soil Sci., v. 29, p. 457-479.

Hendricks, S. B., and Jefferson, M. E., 1938, Crystal structure of vermiculites and mixed vermiculite-chlorites: Am. Mineralogist, v. 23, no. 12, p. 852-862.

Hendricks, S. B., and Ross, C. S., 1941, Chemical composition and genesis of glauconite and celadonite: Am. Mineralogist, v. 26, no. 12, p. 683-709.

Hey, M. H., 1954, New review of the chlorites: Mineralog. Mag., v. 30, p. 277-292.

Hinckley, D. N., 1963, Variability in "crystallinity" values among the kaolin deposits of the Coastal Plain of Georgia and South Carolina, *in* E. Ingerson, *Editor*, Clays and clay minerals: Pergamon Press, New York, p. 229-235.

Hower, J., 1961, Some factors concerning the nature and origin of glauconite: Am. Mineralogist, v. 46, no. 3-4, p. 313-334.

Hower, J., and Mowatt, T. C., 1966, The mineralogy of illites and mixed-layer illite/montmorillonites: Am. Mineralogist, v. 51, no. 5-6, p. 825-854.

Jackson, M. L., Hseung, Y., Corey, R. B., Evans, E. J., and Vanden Heuvel, R. C., 1952, Weathering sequence of clay-size minerals in soils and sediments, Pt. 2, Chemical weathering of layer silicates: Soil Sci. Soc. America Proc., v. 16, no. 1, p. 3-6.

James, H. L., 1966, Chemistry of the iron-rich sedimentary rocks, *in* Data of Geochemistry: 6th ed., U.S. Geol. Survey Prof. Paper 440-W, 60 p.

Jonas, E. C., and Brown, T. E., 1959, Analysis of interlayer mixtures of three clay mineral types by X-ray diffraction: Jour. Sed. Petrology, v. 29, no. 1, p. 77-86.

Keely, W. M., 1966, X-ray diffraction technique for rapid surface area determinations: Norelco Reporter, v. 13, no. 3, p. 90.

Kerr, P. F., and Hamilton, P. E., 1949, Glossary of clay mineral names: Am. Petroleum Inst. Project 49, Clay Mineral Standards Prelim. Rept. 1, 68 p.

Kerr, P. F., Main, M. S., and Hamilton, P. K., 1950, Occurrence and microscopic examination of reference clay minerals: Am. Petroleum Inst. Project 49, Clay Mineral Standards Prelim. Rept. 5, 58 p.

Klug, H. P., and Alexander, L. E., 1962, X-ray diffraction procedures for polycrystalline and amorphous materials: 3d printing, John Wiley & Sons, Inc., New York, 716 p.

Kodama, Hideomi, 1958, Mineralogical study on some pyrophyllites from Japan: Mineralog. Jour., v. 2, p. 236-244.

Lippmann, F., 1954, Über einen keuperton von Zaiserweiher bei Mauldronn: Heidelberger Beitr. Miner. u. Petrog., v. 4, no. 1-2, p. 130-134.

MacEwan, D. M. C., 1949, Some notes on the recording and interpretation of X-ray patterns of soil clays: Jour. Soil Sci., v. 1, no. 1, p. 90-113.

—— 1961, Montmorillonite minerals, *in* G. Brown, *Editor,* The X-ray identification and crystal structures of clay minerals: 2d ed., Mineralog. Soc. London, p. 143-207.

MacEwan, D. M. C., Amil, A. Ruiz, and Brown, G., 1961, Interstratified clay minerals, *in* G. Brown, *Editor,* The X-ray identification and crystal structures of clay minerals: 2d ed., Mineralog. Soc. London, p. 393-445.

MacKenzie, R. C., and Mitchell, B. D., 1966, Clay mineralogy: Earth-Science Rev., v. 2, no. 1, p. 47-91.

Marel, H. W. van der, 1966, Quantitative analysis of clay minerals and their admixtures: Cont. Mineralog. and Petrolog., v. 12, no. 1, p. 96-138.

Millot, G., 1964, Géologie des argiles; alterations, sedimentologie, géochimie: Masson, Paris, 449 p.

Nagy, B. S., and Bradley, W. F., 1955, The structural scheme of sepiolite: Am. Mineralogist, v. 40, no. 9-10, p. 885-892.

Neumann, H., 1956, A pulverizer for mica and micaceous materials: Saertr. Norsk Geologisk Tidsskrift, v. 36, no. 1, p. 52-54.

Newnham, R. E., 1961, A refinement of the dickite structure and some remarks on polymorphism in kaolin minerals: Mineralog. Mag., v. 32, no. 252, p. 683-704.

Newnham, R. E., and Brindley, G. W., 1956, The crystal structure of dickite: Acta Cryst., v. 9, no. 9, p. 759-764.

Norrish, K., and Taylor, R. M., 1962, Quantitative analysis by X-ray diffraction: Clay Minerals Bull., v. 5, no. 28, p. 98-109.

Oinuma, K., and Kobayashi, K., 1961, Problems of rapid clay mineralogical analysis of sedimentary rocks: Clay Sci., v. 1, no. 1-2, p. 8-15.

Porrenga, D. H., 1965, Chamosite in recent sediments of the Niger and Orinoco deltas: Geol. en Mijnbouw, v. 44, no. 11, p. 400-403.

Radoslovich, E. W., and Norrish, K., 1962, The cell dimensions and symmetry of layer-lattice silicates—Pt. 1, Some structural considerations: Am. Mineralogist, v. 47, no. 5-6, p. 599-616.

Reynolds, R. C., 1967, Interstratified clay systems—calculation of the total one-dimensional diffraction function: Am. Mineralogist, v. 52, no. 5-6, p. 661-672.

Rich, C. I., and Obenshain, S. S., 1955, Chemical and clay mineral properties of a red-yellow podzolic soil derived from muscovite schist: Soil Sci. Soc. America Proc., v. 19, no. 3, p. 334-339.

Ross, C. S., and Hendricks, S. B., 1945, Minerals of the montmorillonite group, their origin and relation to soils and clays: U.S. Geol. Survey Prof. Paper 205-B, p. 23-79.

Ross, M., Takeda, H., and Wones, D. R., 1966, Mica polytypes—systematic description and identification: Science, v. 151, no. 2707, p. 191-193.

Schoen, R., 1962, Semiquantitative analysis of chlorites by X-ray diffraction: Am. Mineralogist, v. 47, p. 1384-1392.

Schultz, L. G., 1964, Quantitative interpretation of mineralogical composition from X-ray and chemical data for the Pierre Shale: U.S. Geol. Survey Prof. Paper 391-C, 31 p.

Shirozu, H., 1958, X-ray powder patterns and cell dimensions of some chlorites in Japan, with a note on their interference colors: Mineralog. Jour., (Japan), v. 2, no. 4, p. 209-223.

Siever, R., and Kastner, M., 1967, Mineralogy and petrology of some Mid-Atlantic Ridge sediments: Jour. Marine Research, v. 25, no. 3, p. 263-278.

Smith, J. V., and Yoder, H. S., 1956, Studies of mica polymorphs: Mineralog. Mag., v. 31, p. 209-235.

Stevens, R. E., 1946, A system for calculating analyses of micas and related minerals to end members, *in* Contributions to geochemistry 1942-45: U.S. Geol. Survey Bull. 950, p. 101-119.

Sudo, T., 1959, Mineralogical study on clays of Japan: Maruzon, Tokyo

Sudo, T., Oinuma, K., and Kobayashi, K., 1961, Mineralogical problems concerning rapid clay mineral analysis of sedimentary rocks: Acta Universitatis Carolinae Geol. Supp., 1, p. 189-219.

Switzer, G., Axelrod, J. M., Lindberg, M. L., and Larsen, E. S., III, 1948, Tables of *d* spacings for angle 2 θ CuKα, CuKα_1, CuKα_2, FeKα_1, KeKα_2: U.S. Geol. Survey Circ. 29, 10 p.

Talvenheimo, G., and White, J. L., 1952, Quantitative analysis of clay minerals with the X-ray spectrometer: Anal. Chemistry, v. 24, p. 1784-1789.

Tatlock, D. B., 1966, Rapid modal analysis of some felsic rocks from calibrated X-ray diffraction patterns: U.S. Geol. Survey Bull. 1209, 41 p.

Taylor, J. H., 1949, Petrology of the Northampton sand ironstone formation: Great Britain Geol. Survey Mem., 111 p.

Truesdall, P. E., and Varnes, D. J., 1950, Chart correlating various grain-size definitions of sedimentary materials: U.S. Geol. Survey.

U.S. Department of Agriculture, 1951, Soil survey manual: U.S. Dept. Agric. Handbook No. 18, Soil Survey Staff, Bur. Plant Industry, Soil and Agric. Eng., 502 p.

U.S. National Bureau of Standards, 1950, Tables for conversion of X-ray diffraction angles to interplanar spacing: U.S. Natl. Bur. Standards Applied Mathematics Ser., no. 10, 159 p.

Velde, B., 1965, Experimental determination of muscovite polymorph stabilities: Am. Mineralogist, v. 50, no. 3-4, p. 436-449.

Wada, K., 1967, A structural scheme of soil allophane: Am. Mineralogist, v. 52, no. 5-6, p. 690-708.

Walker, G. F., 1961, The vermiculite minerals, *in* G. Brown, *Editor,* The X-ray identification and crystal structures of clay minerals: 2d ed., Mineralog. Soc. London, p. 297-324.

Warshaw, C. M., 1957, The mineralogy of glauconite: Ph.D. thesis, Penn. State Univ., 167 p.

Weaver, C. E., 1956, The distribution and identification of mixed-layer clays in sedimentary rocks: Am. Mineralogist, v. 41, no. 3-4, p. 202-221.

Willis, A. L., Pennington, R. P., and Jackson, M. L., 1948, Mineral standards for quantitative X-ray diffraction analysis of soil clays; I. Abridgement of component percentages based on weathering sequence: Soil Sci. Soc. America Proc., v. 12, p. 400-416.

Yoder, H. S., and Eugster, H. D., 1955, Synthetic and natural muscovites: Geochim. et Cosmochim. Acta, v. 8, no. 5-6, p. 225-280.

Zen, E-an, 1967, Mixed-layer minerals as one-dimensional crystals: Am. Mineralogist, v. 52, no. 5-6, p. 635-660.

MANUSCRIPT RECEIVED BY THE SOCIETY DECEMBER 5, 1968
REVISED MANUSCRIPT RECEIVED APRIL 1, 1969
PUBLICATION AUTHORIZED BY THE DIRECTOR,
U.S. GEOLOGICAL SURVEY

Author Index

Subject Index